CAMBRIDGE LIBRARY COLLECTION

Books of enduring scholarly value

Darwin

Two hundred years after his birth and 150 years after the publication of 'On the Origin of Species', Charles Darwin and his theories are still the focus of worldwide attention. This series offers not only works by Darwin, but also the writings of his mentors in Cambridge and elsewhere, and a survey of the impassioned scientific, philosophical and theological debates sparked by his 'dangerous idea'.

The Life and Letters of Charles Darwin

This three-volume life of Charles Darwin, published five years after his death, was edited by his son Francis, who was his father's collaborator in experiments in botany and who after his death took on the responsibility of overseeing the publication of his remaining manuscript works and letters. In the preface to the first volume, Francis Darwin explains his editorial principles: 'In choosing letters for publication I have been largely guided by the wish to illustrate my father's personal character. But his life was so essentially one of work, that a history of the man could not be written without following closely the career of the author.' Among the family history, anecdotes and reminiscences of scientific colleagues is a short autobiographical essay which Charles Darwin wrote for his children and grandchildren, rather than for publication. This account of Darwin the man has never been bettered.

Cambridge University Press has long been a pioneer in the reissuing of out-of-print titles from its own backlist, producing digital reprints of books that are still sought after by scholars and students but could not be reprinted economically using traditional technology. The Cambridge Library Collection extends this activity to a wider range of books which are still of importance to researchers and professionals, either for the source material they contain, or as landmarks in the history of their academic discipline.

Drawing from the world-renowned collections in the Cambridge University Library, and guided by the advice of experts in each subject area, Cambridge University Press is using state-of-the-art scanning machines in its own Printing House to capture the content of each book selected for inclusion. The files are processed to give a consistently clear, crisp image, and the books finished to the high quality standard for which the Press is recognised around the world. The latest print-on-demand technology ensures that the books will remain available indefinitely, and that orders for single or multiple copies can quickly be supplied.

The Cambridge Library Collection will bring back to life books of enduring scholarly value across a wide range of disciplines in the humanities and social sciences and in science and technology.

The Life and Letters of Charles Darwin

Including an Autobiographical Chapter

VOLUME 3

CHARLES DARWIN
EDITED BY FRANCIS DARWIN

CAMBRIDGE
UNIVERSITY PRESS

CAMBRIDGE UNIVERSITY PRESS

Cambridge New York Melbourne Madrid Cape Town Singapore São Paolo Delhi

Published in the United States of America by Cambridge University Press, New York

www.cambridge.org
Information on this title: www.cambridge.org/9781108003476

This edition first published 1887
This digitally printed version 2009

ISBN 978-1-108-00347-6

This book reproduces the text of the original edition. The content and language reflect the beliefs, practices and terminology of their time, and have not been updated.

FROM A PHOTOGRAPH (1881) BY MESSRS. ELLIOTT AND FRY.

Frontispiece, Vol. III.

THE LIFE AND LETTERS OF CHARLES DARWIN,

INCLUDING

AN AUTOBIOGRAPHICAL CHAPTER.

EDITED BY HIS SON,

FRANCIS DARWIN.

IN THREE VOLUMES:—Vol. III.

LONDON:
JOHN MURRAY, ALBEMARLE STREET.
1887.

TABLE OF CONTENTS.

VOLUME III.

BOTANICAL LETTERS.

APPENDICES.

ILLUSTRATIONS.

ERRATA.

VOLUME III.

P. 40, line 13: *for* "Magazines" *read* "Magazine."
P. 46, footnote, last line: *for* "contemporaine" *read* "contemporain."
P. 58, line 8: *for* "laburnums, Adami-trifacial" *read* "laburnum Adami, trifacial."

LIFE AND LETTERS

OF

CHARLES DARWIN.

CHAPTER I.

THE SPREAD OF EVOLUTION.

'VARIATION OF ANIMALS AND PLANTS.'

1863–1866.

HIS book on animals and plants under domestication was my father's chief employment in the year 1863. His diary records the length of time spent over the composition of its chapters, and shows the rate at which he arranged and wrote out for printing the observations and deductions of several years.

The three chapters in vol. ii. on inheritance, which occupy 84 pages of print, were begun in January and finished on April 1st; the five on crossing, making 106 pages, were written in eight weeks, while the two chapters on selection, covering 57 pages, were begun on June 16th and finished on July 20th.

The work was more than once interrupted by ill-health, and, in September, what proved to be the beginning of a six months' illness forced him to leave home for the water-cure at Malvern. He returned in October, and remained ill and depressed, in spite of the hopeful opinion of one of the most cheery and skilful physicians of the day. Thus he wrote to Sir J. D. Hooker in November:—

"Dr. Brinton has been here (recommended by Busk); he

does not believe my brain or heart are primarily affected, but I have been so steadily going downhill, I cannot help doubting whether I can ever crawl a little uphill again. Unless I can, enough to work a little, I hope my life may be very short, for to lie on a sofa all day and do nothing but give trouble to the best and kindest of wives and good dear children is dreadful."

The minor works in this year were a short paper in the 'Natural History Review' (N.S. vol. iii. p. 115), entitled "On the so-called *Auditory-Sac* of Cirripedes," and one in the 'Geological Society's Journal' (vol. xix.), on the "Thickness of the Pampæan Formation near Buenos Ayres." The paper on Cirripedes was called forth by the criticisms of a German naturalist Krohn,* and is of some interest in illustration of my father's readiness to admit an error.

With regard to the spread of a belief in Evolution, it could not yet be said that the battle was won, but the growth of belief was undoubtedly rapid. So that, for instance, Charles Kingsley could write to F. D. Maurice :—†

"The state of the scientific mind is most curious; Darwin is conquering everywhere, and rushing in like a flood, by the mere force of truth and fact."

Mr. Huxley was as usual active in guiding and stimulating the growing tendency to tolerate or accept the views set forth in the 'Origin of Species.' He gave a series of lectures to working men at the School of Mines in November, 1862. These were printed in 1863 from the shorthand notes of Mr. May, as six little blue books, price 4*d.* each, under the title, 'Our Knowledge of the Causes of Organic Nature.' When published they were read with interest by my father, who thus refers to them in a letter to Sir J. D. Hooker :—

* Krohn stated that the structures described by my father as ovaries were in reality salivary glands, also that the oviduct runs down to the orifice described in the 'Monograph of the Cirripedia' as the auditory *meatus*.

† Kingsley's 'Life,' vol. ii. p. 171.

"I am very glad you like Huxley's lectures. I have been very much struck with them, especially with the 'Philosophy of Induction.' I have quarrelled with him for overdoing sterility and ignoring cases from Gärtner and Kölreuter about sterile varieties. His geology is obscure; and I rather doubt about man's mind and language. But it seems to me *admirably* done, and, as you say, "Oh my!" about the praise of the 'Origin.' I can't help liking it, which makes me rather ashamed of myself."

My father admired the clearness of exposition shown in the lectures, and in the following letter urges their author to make use of his powers for the advantage of students:]

C. Darwin to T. H. Huxley.

Nov. 5 [1864].

I want to make a suggestion to you, but which may probably have occurred to you. —— was reading your Lectures and ended by saying, "I wish he would write a book." I answered, "he has just written a great book on the skull." "I don't call that a book," she replied, and added, "I want something that people can read; he does write so well." Now, with your ease in writing, and with knowledge at your fingers' ends, do you not think you could write a popular Treatise on Zoology? Of course it would be some waste of time, but I have been asked more than a dozen times to recommend something for a beginner and could only think of Carpenter's Zoology. I am sure that a striking Treatise would do real service to science by educating naturalists. If you were to keep a portfolio open for a couple of years, and throw in slips of paper as subjects crossed your mind, you would soon have a skeleton (and that seems to me the difficulty) on which to put the flesh and colours in your inimitable manner. I believe such a book might have a brilliant success, but I did not intend to scribble so much about it.

Give my kindest remembrance to Mrs. Huxley, and tell

her I was looking at 'Enoch Arden,' and as I know how she admires Tennyson, I must call her attention to two sweetly pretty lines (p. 105) . . .

> . . . and he meant, he said he meant,
> Perhaps he meant, or partly meant, you well.

Such a gem as this is enough to make me young again, and like poetry with pristine fervour.

My dear Huxley,
Yours affectionately,
CH. DARWIN.

[In another letter (Jan. 1865) he returns to the above suggestion, though he was in general strongly opposed to men of science giving up to the writing of text-books, or to teaching, the time that might otherwise have been given to original research.

"I knew there was very little chance of your having time to write a popular treatise on Zoology, but you are about the one man who could do it. At the time I felt it would be almost a sin for you to do it, as it would of course destroy some original work. On the other hand I sometimes think that general and popular treatises are almost as important for the progress of science as original work."

The series of letters will continue the history of the year 1863.]

C. Darwin to J. D. Hooker.

Down, Jan. 3 [1863].

MY DEAR HOOKER.—I am burning with indignation and must exhale. . . . I could not get to sleep till past 3 last night for indignation.* . . .

* It would serve no useful purpose if I were to go into the matter which so strongly roused my father's anger. It was a question of literary dishonesty, in which a friend was the sufferer, but which in no way affected himself.

Now for pleasanter subjects; we were all amused at your defence of stamp collecting and collecting generally. . . . But, by Jove, I can hardly stomach a grown man collecting stamps. Who would ever have thought of your collecting Wedgwood-ware! but that is wholly different, like engravings or pictures. We are degenerate descendants of old Josiah W., for we have not a bit of pretty ware in the house.

. . . Notwithstanding the very pleasant reason you give for our not enjoying a holiday, namely, that we have no vices, it is a horrid bore. I have been trying for health's sake to be idle, with no success. What I shall now have to do, will be to erect a tablet in Down Church, "Sacred to the Memory, &c.," and officially die, and then publish books, "by the late Charles Darwin," for I cannot think what has come over me of late; I always suffered from the excitement of talking, but now it has become ludicrous. I talked lately 1½ hours (broken by tea by myself) with my nephew, and I was [ill] half the night. It is a fearful evil for self and family.

Good-night. Ever yours,

C. DARWIN.

[The following letter to Sir Julius von Haast,* is an example of the sympathy which he felt with the spread and growth of science in the colonies. It was a feeling not expressed once only, but was frequently present in his mind, and often found utterance. When we, at Cambridge, had the satisfaction of receiving Sir J. von Haast into our body as a Doctor of Science (July 1886), I had the opportunity of hearing from him of the vivid pleasure which this, and other letters from my father, gave him. It was pleasant to see how strong had been the impression made by my father's warm-hearted sympathy—an impression which seemed,

* The late Sir Julius von Haast was a German by birth, but had long been resident in New Zealand. He was, in 1862, Government Geologist to the Province of Canterbury.

after more than twenty years, to be as fresh as when it was first received :]

C. Darwin to Julius von Haast.

Down, Jan. 22 [1863].

DEAR SIR,—I thank you most sincerely for sending me your Address and the Geological Report.* I have seldom in my life read anything more spirited and interesting than your address. The progress of your colony makes one proud, and it is really admirable to see a scientific institution founded in so young a nation. I thank you for the very honourable notice of my 'Origin of Species.' You will easily believe how much I have been interested by your striking facts on the old glacial period, and I suppose the world might be searched in vain for so grand a display of terraces. You have, indeed, a noble field for scientific research and discovery. I have been extremely much interested by what you say about the tracks of supposed [living] mammalia. Might I ask, if you succeed in discovering what the creatures are, you would have the great kindness to inform me? Perhaps they may turn out something like the Solenhofen bird creature, with its long tail and fingers, with claws to its wings! I may mention that in South America, in completely uninhabited regions, I found spring rat-traps, baited with *cheese*, were very successful in catching the smaller mammals. I would venture to suggest to you to urge on some of the capable members of your institution to observe annually the rate and manner of spreading of European weeds and insects, and especially to observe *what native plants most fail;* this latter point has never been attended to. Do the introduced hive-bees replace any other insect? &c. All such points are, in my opinion, great desiderata in

* Address to the 'Philosophical Institute of Canterbury (N.Z.).' The "Report" is given in the *New Zealand Government Gazette, Province of Canterbury*, Oct. 1862.

science. What an interesting discovery that of the remains of prehistoric man!

Believe me, dear Sir,

With the most cordial respect and thanks,

Yours very faithfully,

CHARLES DARWIN.

*C. Darwin to Camille Dareste.**

Down, Feb. 16 [1863].

DEAR AND RESPECTED SIR.—I thank you sincerely for your letter and your pamphlet. I had heard (I think in one of M. Quatrefages' books) of your work, and was most anxious to read it, but did not know where to find it. You could not have made me a more valuable present. I have only just returned home, and have not yet read your work; when I do if I wish to ask any questions I will venture to trouble you. Your approbation of my book on Species has gratified me extremely. Several naturalists in England, North America, and Germany, have declared that their opinions on the subject have in some degree been modified, but as far as I know, my book has produced no effect whatever in France, and this makes me the more gratified by your very kind expression of approbation. Pray believe me, dear Sir, with much respect,

Yours faithfully and obliged,

CH. DARWIN.

C. Darwin to J. D. Hooker.

Down, Feb. 24 [1863].

MY DEAR HOOKER.—I am astonished at your note. I have

* Professor Dareste is a well-known worker in Animal Teratology. He was in 1863 living at Lille, but has since then been called to Paris. My father took a special interest in Dareste's work on the production of monsters, as bearing on the causes of variation.

not seen the *Athenæum*,* but I have sent for it, and may get it to-morrow; and will then say what I think.

I have read Lyell's book. ['The Antiquity of Man.'] The whole certainly struck me as a compilation, but of the highest class, for when possible the facts have been verified on the spot, making it almost an original work. The Glacial chapters seem to me best, and in parts magnificent. I could hardly judge about Man, as all the gloss of novelty was completely worn off. But certainly the aggregation of the evidence produced a very striking effect on my mind. The chapter comparing language and changes of species, seems most ingenious and interesting. He has shown great skill in picking out salient points in the argument for change of species; but I am deeply disappointed (I do not mean personally) to find that his timidity prevents him giving any judgment. . . . From all my communications with him I must ever think that he has really entirely lost faith in the immutability of species; and yet one of his strongest sentences is nearly as follows: "If it should *ever*† be rendered highly probable that species change by variation and natural selection," &c. &c. I had hoped he would have guided the public as far as his own belief went. . . . One thing does please me on this subject, that he seems to appreciate your work. No doubt the public or a part may be induced to think that, as he gives to us a larger space than to Lamarck, he must think there is something in our views. When reading the brain chapter, it struck me forcibly that if

* In the 'Antiquity of Man,' first edition, p. 480, Lyell criticised somewhat severely Owen's account of the difference between the Human and Simian brains. The number of the *Athenæum* here referred to (1863, p. 262) contains a reply by Professor Owen to Lyell's strictures. The surprise expressed by my father was at the revival of a controversy which every one believed to be closed. Prof. Huxley (*Medical Times*, Oct. 25, 1862, quoted in 'Man's Place in Nature,' p. 117) spoke of the "two years during which this preposterous controversy has dragged its weary length." And this no doubt expressed a very general feeling.

† The italics are not Lyell's.

he had said openly that he believed in change of species, and as a consequence that man was derived from some Quadrumanous animal, it would have been very proper to have discussed by compilation the differences in the most important organ, viz. the brain. As it is, the chapter seems to me to come in rather by the head and shoulders. I do not think (but then I am as prejudiced as Falconer and Huxley, or more so) that it is too severe; it struck me as given with judicial force. It might perhaps be said with truth that he had no business to judge on a subject on which he knows nothing; but compilers must do this to a certain extent. (You know I value and rank high compilers, being one myself!) I have taken you at your word, and scribbled at great length. If I get the *Athenæum* to-morrow, I will add my impression of Owen's letter.

. . . The Lyells are coming here on Sunday evening to stay till Wednesday. I dread it, but I must say how much disappointed I am that he has not spoken out on species, still less on man. And the best of the joke is that he thinks he has acted with the courage of a martyr of old. I hope I may have taken an exaggerated view of his timidity, and shall *particularly* be glad of your opinion on this head.* When I got his book I turned over the pages, and saw he had discussed the subject of species, and said that I thought he would do more to convert the public than all of us, and now (which makes the case worse for me) I must, in common honesty, retract. I wish to Heaven he had said not a word on the subject.

Wednesday morning: I have read the *Athenæum*. I do not think Lyell will be nearly so much annoyed as you expect. The concluding sentence is no doubt very stinging.

* On this subject my father wrote to Sir Joseph Hooker: "Cordial thanks for your deeply interesting letters about Lyell, Owen, and Co. I cannot say how glad I am to hear that I have not been unjust about the species-question towards Lyell. I feared I had been unreasonable."

No one but a good anatomist could unravel Owen's letter; at least it is quite beyond me.

. . . Lyell's memory plays him false when he says all anatomists were astonished at Owen's paper;* it was often quoted with approbation. I *well* remember Lyell's admiration at this new classification! (Do not repeat this.) I remember it, because, though I knew nothing whatever about the brain, I felt a conviction that a classification thus founded on a single character would break down, and it seemed to me a great error not to separate more completely the Marsupialia. . . .

What an accursed evil it is that there should be all this quarrelling within, what ought to be, the peaceful realms of science. I will go to my own present subject of inheritance and forget it all for a time. Farewell, my dear old friend,

C. DARWIN.

C. Darwin to Asa Gray.

Down, Feb. 23 [1863].

. . . If you have time to read you will be interested by parts of Lyell's book on man; but I fear that the best part, about the Glacial period, may be too geological for any one except a regular geologist. He quotes you at the end with gusto. By the way, he told me the other day how pleased some had been by hearing that they could purchase your pamphlet. The *Parthenon* also speaks of it as the ablest contribution to the literature of the subject. It delights me when I see your work appreciated.

The Lyells come here this day week, and I shall grumble at his excessive caution. . . . The public may well say, if such a man dare not or will not speak out his mind, how can we who are ignorant form even a guess on the subject? Lyell was pleased when I told him lately that you thought that language might be used as an excellent illustration of deriva-

* "On the Characters, &c., of the Class Mammalia," 'Linn. Soc. Journal,' ii. 1858.

tion of species; you will see that he has an *admirable* chapter on this. . . .

I read Cairns's excellent Lecture,* which shows so well how your quarrel arose from Slavery. It made me for a time wish honestly for the North; but I could never help, though I tried, all the time thinking how we should be bullied and forced into a war by you, when you were triumphant. But I do most truly think it dreadful that the South, with its accursed slavery, should triumph, and spread the evil. I think if I had power, which, thank God, I have not, I would let you conquer the border States, and all west of the Mississippi, and then force you to acknowledge the cotton States. For do you not now begin to doubt whether you can conquer and hold them? I have inflicted a long tirade on you.

The *Times* is getting more detestable (but that is too weak a word) than ever. My good wife wishes to give it up, but I tell her that is a pitch of heroism to which only a woman is equal. To give up the "Bloody Old *Times*," as Cobbett used to call it, would be to give up meat, drink and air. Farewell, my dear Gray,

Yours most truly,
C. DARWIN.

C. Darwin to C. Lyell.

Down, March 6, [1863].

. . . I have been of course deeply interested by your book.† I have hardly any remarks worth sending, but will scribble a little on what most interested me. But I will first get out what I hate saying, viz. that I have been greatly disappointed that you have not given judgment and spoken fairly out what you think about the derivation of species. I should have been contented if you had boldly said that species have not

* Prof. J. E. Cairns, 'The Slave Power, &c.: an attempt to explain the real issues involved in the American contest.' 1862.

† 'Antiquity of Man.'

been separately created, and had thrown as much doubt as you like on how far variation and natural selection suffices. I hope to Heaven I am wrong (and from what you say about Whewell it seems so), but I cannot see how your chapters can do more good than an extraordinary able review. I think the *Parthenon* is right, that you will leave the public in a fog. No doubt they may infer that as you give more space to myself, Wallace, and Hooker, than to Lamarck, you think more of us. But I had always thought that your judgment would have been an epoch in the subject. All that is over with me, and I will only think on the admirable skill with which you have selected the striking points, and explained them. No praise can be too strong, in my opinion, for the inimitable chapter on language in comparison with species.

p. 505—A sentence * at the top of the page makes me groan. . . .

I know you will forgive me for writing with perfect freedom, for you must know how deeply I respect you as my old honoured guide and master. I heartily hope and expect that your book will have gigantic circulation and may do in many ways as much good as it ought to do. I am tired, so no more. I have written so briefly that you will have to guess my meaning. I fear my remarks are hardly worth sending. Farewell, with kindest remembrance to Lady Lyell.

Ever yours,

C. DARWIN.

[Mr. Huxley has quoted (Vol. II. p. 193) some passages from Lyell's letters which show his state of mind at this time. The following passage, from a letter of March 11th to my father, is also of much interest:—

* After speculating on the sudden appearance of individuals far above the average of the human race, Lyell asks if such leaps upwards in the scale of intellect may not "have cleared at one bound the space which separated the highest stage of the unprogressive intelligence of the inferior animals from the first and lowest form of improvable reason manifested by man."

"My feelings, however, more than any thought about policy or expediency, prevent me from dogmatising as to the descent of man from the brutes, which, though I am prepared to accept it, takes away much of the charm from my speculations on the past relating to such matters. . . . But you ought to be satisfied, as I shall bring hundreds towards you who, if I treated the matter more dogmatically, would have rebelled."]

C. Darwin to C. Lyell.

Down, 12th [March, 1863].

MY DEAR LYELL,—I thank you for your very interesting and kind, I may say, charming letter. I feared you might be huffed for a little time with me. I know some men would have been so. I have hardly any more criticisms, anyhow, worth writing. But I may mention that I felt a little surprise that old B. de Perthes * was not rather more honourably mentioned. I would suggest whether you could not leave out some references to the 'Principles;' one for the real student is as good as a hundred, and it is rather irritating, and gives a feeling of incompleteness to the general reader to be often referred to other books. As you say that you have gone as far as you believe on the species question, I have not a word to say; but I must feel convinced that at times, judging from conversation, expressions, letters, &c., you have as completely given up belief in immutability of specific forms as I have done. I must still think a clear expression from you, *if you could have given it*, would have been potent with the public, and all the more so, as you formerly held opposite opinions. The more I work, the more satisfied I become with variation and natural selection, but that part of the case I look at as less important, though more interesting to me personally. As you ask for criticisms on this head (and believe me that

* Born 1788, died 1868. See footnote, p. 16.

I should not have made them unasked), I may specify (pp. 412, 413) that such words as "Mr. D. labours to show," "is believed by the author to throw light," would lead a common reader to think that you yourself do *not* at all agree, but merely think it fair to give my opinion. Lastly, you refer repeatedly to my view as a modification of Lamarck's doctrine of development and progression. If this is your deliberate opinion there is nothing to be said, but it does not seem so to me. Plato, Buffon, my grandfather before Lamarck, and others, propounded the *obvious* view that if species were not created separately they must have descended from other species, and I can see nothing else in common between the 'Origin' and Lamarck. I believe this way of putting the case is very injurious to its acceptance, as it implies necessary progression, and closely connects Wallace's and my views with what I consider, after two deliberate readings, as a wretched book, and one from which (I well remember my surprise) I gained nothing. But I know you rank it higher, which is curious, as it did not in the least shake your belief. But enough, and more than enough. Please remember you have brought it all down on yourself!!

I am very sorry to hear about Falconer's "reclamation." * I hate the very word, and have a sincere affection for him.

Did you ever read anything so wretched as the *Athenæum* reviews of you, and of Huxley † especially. Your *object* to make man old, and Huxley's *object* to degrade him. The wretched writer has not a glimpse what the discovery of scientific truth means. How splendid some pages are in Huxley, but I fear the book will not be popular. . . .

* "Falconer, whom I referred to oftener than to any other author, says I have not done justice to the part he took in resuscitating the cave question, and says he shall come out with a separate paper to prove it. I offered to alter anything in the new edition, but this he declined."—C. Lyell to C. Darwin, March 11, 1863; Lyell's 'Life,' vol. ii. p. 364.

† 'Man's Place in Nature,' 1863.

C. Darwin to J. D. Hooker.

Down [March 13, 1863].

I should have thanked you sooner for the *Athenæum* and very pleasant previous note, but I have been busy, and not a little uncomfortable from frequent uneasy feeling of fullness, slight pain and tickling about the heart. But as I have no other symptoms of heart complaint I do not suppose it is affected. . . . I have had a most kind and delightfully candid letter from Lyell, who says he spoke out as far as he believes. I have no doubt his belief failed him as he wrote, for I feel sure that at times he no more believed in Creation than you or I. I have grumbled a bit in my answer to him at his *always* classing my work as a modification of Lamarck's, which it is no more than any author who did not believe in immutability of species, and did believe in descent. I am very sorry to hear from Lyell that Falconer is going to publish a formal reclamation of his own claims. . . .

It is cruel to think of it, but we must go to Malvern in the middle of April; it is ruin to me.* . . .

C. Darwin to C. Lyell.

Down, March 17 [1863].

MY DEAR LYELL,—I have been much interested by your letters and enclosure, and thank you sincerely for giving me so much time when you must be so busy. What a curious letter from B. de P. [Boucher de Perthes]. He seems perfectly satisfied, and must be a very amiable man. I know something about his errors, and looked at his book many years ago, and am ashamed to think that I concluded the

* He went to Hartfield, in Sussex, on April 27, and to Malvern in the autumn.

whole was rubbish! Yet he has done for man something like what Agassiz did for glaciers.*

I cannot say that I agree with Hooker about the public not liking to be told what to conclude, *if coming from one in your position.* But I am heartily sorry that I was led to make complaints, or something very like complaints, on the manner in which you have treated the subject, and still more so anything about myself. I steadily *endeavour* never to forget my firm belief that no one can at all judge about his own work. As for Lamarck, as you have such a man as Grove with you, you are triumphant; not that I can alter my opinion that to me it was an absolutely useless book. Perhaps this was owing to my always searching books for facts, perhaps from knowing my grandfather's earlier and identically the same speculation. I will only further say that if I can analyse my own feelings (a very doubtful process), it is nearly as much for your sake as for my own, that I so much wish that your state of belief could have permitted you to say boldly and distinctly out that species were not separately created. I have generally told you the progress of opinion, as I have heard it, on the species question. A first-rate German naturalist † (I now forget the name!), who has lately published a grand folio, has spoken out to the utmost extent on the 'Origin.' De Candolle, in a very good paper on "Oaks," goes, in Asa Gray's opinion, as far as he himself does; but De Candolle, in writing to me, says *we*, "we think this and that;" so that I infer he really goes to the full extent with me, and tells me of a French good botanical palæontologist (name

* In his 'Antiquités Celtiques' (1847), Boucher de Perthes described the flint tools found at Abbeville with bones of rhinoceros, hyæna, &c. "But the scientific world had no faith in the statement that works of art, however rude, had been met with in undisturbed beds of such antiquity." ('Antiquity of Man,' first edition, p. 95.)

† No doubt Haeckel, whose monograph on the Radiolaria was published in 1862. In the same year Professor W. Preyer of Jena published a Dissertation on *Alca impennis*, which was one of the earliest pieces of special work on the basis of the 'Origin of Species.'

forgotten),* who writes to De Candolle that he is sure that my views will ultimately prevail. But I did not intend to have written all this. It satisfies me with the final results, but this result, I begin to see, will take two or three life-times. The entomologists are enough to keep the subject back for half a century. I really pity your having to balance the claims of so many eager aspirants for notice; it is clearly impossible to satisfy all. . . . Certainly I was struck with the full and due honour you conferred on Falconer. I have just had a note from Hooker. . . . I am heartily glad that you have made him so conspicuous; he is so honest, so candid, and so modest. . . .

I have read ——. I could find nothing to lay hold of, which in one sense I am very glad of, as I should hate a controversy; but in another sense I am very sorry for, as I long to be in the same boat with all my friends. . . . I am heartily glad the book is going off so well.

Ever yours,

C. DARWIN.

C. Darwin to J. D. Hooker.

Down [March 29, 1863].

. . . Many thanks for *Athenæum*, received this morning, and to be returned to-morrow morning. Who would have ever thought of the old stupid *Athenæum* taking to Oken-like transcendental philosophy written in Owenian style!† . . . .

* The Marquis de Saporta.

† This refers to a review of Dr. Carpenter's 'Introduction to the study of Foraminifera,' that appeared in the *Athenæum* of March 28, 1863 (p. 417). The reviewer attacks Dr. Carpenter's views in as much as they support the doctrine of Descent; and he upholds spontaneous generation (Heterogeny) in place of what Dr. Carpenter, naturally enough, believed in, viz. the genetic connection of living and extinct Foraminifera. In the next number is a letter by Dr. Carpenter, which chiefly consists of a protest against the reviewer's somewhat contemptuous classification of Dr. Carpenter and my father as disciple and master. In the course of the letter Dr. Carpenter says—p. 461:—

It will be some time before we see "slime, protoplasm, &c." generating a new animal.* But I have long regretted that I truckled to public opinion, and used the Pentateuchal term of creation,† by which I really meant "appeared" by some wholly unknown process. It is mere rubbish, thinking at present of the origin of life; one might as well think of the origin of matter.

C. Darwin to J. D. Hooker.

Down, Friday night [April 17, 1863].

MY DEAR HOOKER,—I have heard from Oliver that you will be now at Kew, and so I am going to amuse myself by scribbling a bit. I hope you have thoroughly enjoyed your

---

"Under the influence of his foregone conclusion that I have accepted Mr. Darwin as my master, and his hypothesis as my guide, your reviewer represents me as blind to the significance of the general fact stated by me, that 'there has been no advance in the foraminiferous type from the palæozoic period to the present time.' But for such a foregone conclusion he would have recognised in this statement the expression of my conviction that the present state of scientific evidence, instead of sanctioning the idea that the descendants of the primitive type or types of Foraminifera can ever rise to any higher grade, justifies the *anti-Darwinian* inference, that however widely they diverge from each other and from their originals, *they still remain Foraminifera*."

* On the same subject my father wrote in 1871: "It is often said that all the conditions for the first production of a living organism are now present, which could ever have been present. But if (and oh! what a big if!) we could conceive in some warm little pond, with all sorts of ammonia and phosphoric salts, light, heat, electricity, &c., present, that a proteine compound was chemically formed ready to undergo still more complex changes, at the present day such matter would be instantly devoured or absorbed, which would not have been the case before living creatures were formed."

† This refers to a passage in which the reviewer of Dr. Carpenter's book speaks of "an operation of force," or "a concurrence of forces which have now no place in nature," as being, "a creative force, in fact, which Darwin could only express in Pentateuchal terms as the primordial form 'into which life was first breathed.'" The conception of expressing a creative force as a primordial form is the Reviewer's.

tour. I never in my life saw anything like the spring flowers this year. What a lot of interesting things have been lately published. I liked extremely your review of De Candolle. What an awfully severe article that by Falconer on Lyell;* I am very sorry for it; I think Falconer on his side does not do justice to old Perthes and Schmerling. . . . . I shall be very curious to see how he [Lyell] answers it to-morrow. (I have been compelled to take in the *Athenæum* for a while.) I am very sorry that Falconer should have written so spitefully, even if there is some truth in his accusations; I was rather disappointed in Carpenter's letter, no one could have given a better answer, but the chief object of his letter seems to me to be to show that though he has touched pitch he is not defiled. No one would suppose he went so far as to believe all birds came from one progenitor. I have written a letter to the *Athenæum* † (the first and last time I shall take such a step)

* *Athenæum*, April 4, 1863, p. 459. The writer asserts that justice has not been done either to himself or Mr. Prestwich—that Lyell has not made it clear that it was their original work which supplied certain material for the 'Antiquity of Man.' Falconer attempts to draw an unjust distinction between a "philosopher" (here used as a polite word for compiler) like Sir Charles Lyell, and original observers, presumably such as himself and Mr. Prestwich. Lyell's reply was published in the *Athenæum*, April 18, 1863. It ought to be mentioned that a letter from Mr. Prestwich (*Athenæum*, p. 555), which formed part of the controversy, though of the nature of a reclamation, was written in a very different spirit and tone from Dr. Falconer's.

† *Athenæum*, 1863, p. 554: "The view given by me on the origin or derivation of species, whatever its weaknesses may be, connects (as has been candidly admitted by some of its opponents. such as Pictet, Bronn, &c.), by an intelligible thread of reasoning, a multitude of facts: such as the formation of domestic races by man's selection,—the classification and affinities of all organic beings,—the innumerable gradations in structure and instincts,—the similarity of pattern in the hand, wing, or paddle of animals of the same great class,—the existence of organs become rudimentary by disuse,—the similarity of an embryonic reptile, bird and mammal, with the retention of traces of an apparatus fitted for aquatic respiration; the retention in the young calf of incisor teeth in the upper jaw, &c.—the distribution of animals and plants, and their mutual affinities within the same region, — their

to say, under the cloak of attacking Heterogeny, a word in my own defence. My letter is to appear next week, so the Editor says; and I mean to quote Lyell's sentence * in his second edition, on the principle if one puffs oneself, one had better puff handsomely. . . .

C. Darwin to C. Lyell.

Down, April 18 [1863].

MY DEAR LYELL,—I was really quite sorry that you had sent me a second copy † of your valuable book. But after a few hours my sorrow vanished for this reason: I have written a letter to the *Athenæum*, in order, under the cloak of attacking the monstrous article on Heterogeny, to say a word for myself in answer to Carpenter, and now I have inserted a few sentences in allusion to your analogous objection ‡ about

---

general geological succession, and the close relationship of the fossils in closely consecutive formations and within the same country; extinct marsupials having preceded living marsupials in Australia, and armadillo-like animals having preceded and generated armadilloes in South America,—and many other phenomena, such as the gradual extinction of old forms and their gradual replacement by new forms better fitted for their new conditions in the struggle for life. When the advocate of Heterogeny can thus connect large classes of facts, and not until then, he will have respectful and patient listeners."

* See the next letter.

† The second edit. of the 'Antiquity of Man' was published a few months after the first had appeared.

‡ Lyell objected that the mammalia (e.g. bats and seals) which alone have been able to reach oceanic islands ought to have become modified into various terrestrial forms fitted to fill various places in their new homes. My father pointed out in the *Athenæum* that Sir Charles has in some measure answered his own objection, and went on to quote the "amended sentence" ('Antiquity of Man,' 2nd edit. p. 469) as showing how far Lyell agreed with the general doctrines of the 'Origin of Species': "Yet we ought by no means to undervalue the importance of the step which will have been made, should it hereafter become the generally received opinion of men of science (as I fully expect it will) that the past changes of the organic world have been brought about by the subordinate agency of such causes as Variation and Natural Selection." In the first edition the words "as I fully expect it will," do not occur.

bats on islands, and then with infinite slyness have quoted your amended sentence, with your parenthesis ("as I fully believe") * ; I do not think you can be annoyed at my doing this, and you see, that I am determined as far as I can, that the public shall see how far you go. This is the first time I have ever said a word for myself in any journal, and it shall, I think, be the last. My letter is short, and no great things. I was extremely concerned to see Falconer's disrespectful and virulent letter. I like extremely your answer just read ; you take a lofty and dignified position, to which you are so well entitled.†

I suspect that if you had inserted a few more superlatives in speaking of the several authors there would have been none of this horrid noise. No one, I am sure, who knows you could doubt about your hearty sympathy with every one who makes any little advance in science. I still well remember my surprise at the manner in which you listened to me in Hart Street on my return from the *Beagle's* voyage. You did me a world of good. It is horridly vexatious that so frank and apparently amiable a man as Falconer should have behaved so.‡ Well, it will all soon be forgotten. . . . .

[In reply to the above-mentioned letter of my father's to the *Athenæum*, an article appeared in that Journal (May 2nd, 1863, p. 586), accusing my father of claiming for his views the exclusive merit of "connecting by an intelligible thread of reasoning" a number of facts in morphology, &c. The writer remarks that, "The different generalisations cited by Mr. Darwin as being connected by an intelligible thread of reasoning exclusively through his

* My father here quotes Lyell incorrectly ; see the footnote on the previous page.

† In a letter to Sir J. D. Hooker he wrote : "I much like Lyell's letter. But all this squabbling will greatly sink scientific men. I have seen sneers already in the *Times*."

‡ It is to this affair that the extract from a letter to Falconer, given Vol. I. p. 158, refers.

attempt to explain specific transmutation are in fact related to it in this wise, that they have prepared the minds of naturalists for a better reception of such attempts to explain the way of the origin of species from species."

To this my father replied as follows in the *Athenæum* of May 9th, 1863 :]

Down, May 5 [1863].

I hope that you will grant me space to own that your reviewer is quite correct when he states that any theory of descent will connect, "by an intelligible thread of reasoning," the several generalizations before specified. I ought to have made this admission expressly; with the reservation, however, that, as far as I can judge, no theory so well explains or connects these several generalizations (more especially the formation of domestic races in comparison with natural species, the principles of classification, embryonic resemblance, &c.) as the theory, or hypothesis, or guess, if the reviewer so likes to call it, of Natural Selection. Nor has any other satisfactory explanation been ever offered of the almost perfect adaptation of all organic beings to each other, and to their physical conditions of life. Whether the naturalist believes in the views given by Lamarck, by Geoffroy St. Hilaire, by the author of the 'Vestiges,' by Mr. Wallace and myself, or in any other such view, signifies extremely little in comparison with the admission that species have descended from other species, and have not been created immutable; for he who admits this as a great truth has a wide field opened to him for further inquiry. I believe, however, from what I see of the progress of opinion on the Continent, and in this country, that the theory of Natural Selection will ultimately be adopted, with, no doubt, many subordinate modifications and improvements.

CHARLES DARWIN.

[In the following, he refers to the above letter to the *Athenæum*:]

C. Darwin to J. D. Hooker.

Leith Hill Place,
Saturday [May 11, 1863].

My dear Hooker,—You give good advice about not writing in newspapers; I have been gnashing my teeth at my own folly; and this not caused by ——'s sneers, which were so good that I almost enjoyed them. I have written once again to own to a certain extent of truth in what he says, and then if I am ever such a fool again, have no mercy on me. I have read the squib in *Public Opinion*;* it is capital; if there is more, and you have a copy, do lend it. It shows well that a scientific man had better be trampled in dirt than squabble. I have been drawing diagrams, dissecting shoots, and muddling my brains to a hopeless degree about the divergence of leaves, and have of course utterly failed. But I can see that the subject is most curious, and indeed astonishing. . . . .

[The next letter refers to Mr. Bentham's presidential

* *Public Opinion*, April 23, 1863. A lively account of a police case, in which the quarrels of scientific men are satirised. Mr. John Bull gives evidence that—

"The whole neighbourhood was unsettled by their disputes; Huxley quarrelled with Owen, Owen with Darwin, Lyell with Owen, Falconer and Prestwich with Lyell, and Gray the menagerie man with everybody. He had pleasure, however, in stating that Darwin was the quietest of the set. They were always picking bones with each other and fighting over their gains. If either of the gravel sifters or stone breakers found anything, he was obliged to conceal it immediately, or one of the old bone collectors would be sure to appropriate it first and deny the theft afterwards, and the consequent wrangling and disputes were as endless as they were wearisome.

"Lord Mayor.—Probably the clergyman of the parish might exert some influence over them?

"The gentleman smiled, shook his head, and stated that he regretted to say that no class of men paid so little attention to the opinions of the clergy as that to which these unhappy men belonged."

address to the Linnean Society (May 25, 1863). Mr. Bentham does not yield to the new theory of Evolution, "cannot surrender at discretion so long as many important outworks remain contestable." But he shows that the great body of scientific opinion is flowing in the direction of belief.

The mention of Pasteur by Mr. Bentham is in reference to the promulgation "as it were *ex cathedrâ*," of a theory of spontaneous generation by the reviewer of Dr. Carpenter in the *Athenæum* (March 28, 1863). Mr. Bentham points out that in ignoring Pasteur's refutation of the supposed facts of spontaneous generation, the writer fails to act with "that impartiality which every reviewer is supposed to possess."]

C. Darwin to G. Bentham.

Down, May 22 [1863].

MY DEAR BENTHAM.—I am much obliged for your kind and interesting letter. I have no fear of anything that a man like you will say annoying me in the very least degree. On the other hand, any approval from one whose judgment and knowledge I have for many years so sincerely respected, will gratify me much. The objection which you well put, of certain forms remaining unaltered through long time and space, is no doubt formidable in appearance, and to a certain extent in reality according to my judgment. But does not the difficulty rest much on our silently assuming that we know more than we do? I have literally found nothing so difficult as to try and always remember our ignorance. I am never weary, when walking in any new adjoining district or country, of reflecting how absolutely ignorant we are why certain old plants are not there present, and other new ones are, and others in different proportions. If we once fully feel this, then in judging the theory of Natural Selection, which implies that a form will remain unaltered unless some alteration be to its benefit, is it so very wonderful that some forms should change much slower and much less, and some few should have changed not at all under conditions which to us (who really know nothing what are the important conditions) seem very different. Certainly *a priori* we might have anticipated that all the plants anciently introduced into Australia would have undergone some modification; but the fact that they have not been modified does not seem to me a difficulty of weight enough to shake a belief grounded on other arguments. I have expressed myself miserably, but I am far from well to-day.

I am very glad that you are going to allude to Pasteur; I was struck with infinite admiration at his work. With cordial thanks, believe me, dear Bentham,

Yours very sincerely,

CH. DARWIN.

P.S.—In fact the belief in Natural Selection must at present be grounded entirely on general considerations. (1) On its being a *vera causa*, from the struggle for existence; and the certain geological fact that species do somehow change. (2) From the analogy of change under domestication by man's selection. (3) And chiefly from this view connecting under an intelligible point of view a host of facts. When we descend to details, we can prove that no one species has changed [*i.e.* we cannot prove that a single species has changed]; nor can we prove that the supposed changes are beneficial, which is the groundwork of the theory. Nor can we explain why some species have changed and others have not. The latter case seems to me hardly more difficult to understand precisely and in detail than the former case of supposed change. Bronn may ask in vain, the old creationist school and the new school, why one mouse has longer ears than another mouse, and one plant more pointed leaves than another plant.

C. Darwin to G. Bentham.

Down, June 19 [1863].

MY DEAR BENTHAM,—I have been extremely much pleased and interested by your address, which you kindly sent me. It seems to be excellently done, with as much judicial calmness and impartiality as the Lord Chancellor could have shown. But whether the "immutable" gentlemen would agree with the impartiality may be doubted, there is too much kindness shown towards me, Hooker, and others, they might say. Moreover I verily believe that your address, written as it is, will do more to shake the unshaken and bring on those leaning to our side, than anything written directly in favour of transmutation. I can hardly tell why it is, but your address has pleased me as much as Lyell's book disappointed me, that is, the part on species, though so cleverly written. I agree with all your remarks on the reviewers. By the way, Lecoq* is a believer in the change of species. I, for one, can conscientiously declare that I never feel surprised at any one sticking to the belief of immutability; though I am often not a little surprised at the arguments advanced on this side. I remember too well my endless oscillations of doubt and difficulty. It is to me really laughable, when I think of the years which elapsed before I saw what I believe to be the explanation of some parts of the case; I believe it was fifteen years after I began before I saw the meaning and cause of the divergence of the descendants of any one pair. You pay me some most elegant and pleasing compliments. There is much in your address which has pleased me much, especially your remarks on various naturalists. I am so glad that you have alluded so honourably to Pasteur. I have just read over this note; it does not express strongly enough the interest which I have felt in reading your address. You have done, I

* Author of 'Géographie Botanique.' 9 vols. 1854–58.

believe, a real good turn to the *right side.* Believe me, dear Bentham,

Yours very sincerely,
CH. DARWIN.

1864.

[In my father's diary for 1864 is the entry, "Ill all January, February, March." About the middle of April (seven months after the beginning of the illness in the previous autumn) his health took a turn for the better. As soon as he was able to do any work, he began to write his papers on Lythrum, and on Climbing Plants, so that the work which now concerns us did not begin until September, when he again set to work on 'Animals and Plants.' A letter to Sir J. D. Hooker gives some account of the re-commencement of the work: "I have begun looking over my old MS., and it is as fresh as if I had never written it; parts are astonishingly dull, but yet worth printing, I think; and other parts strike me as very good. I am a complete millionaire in odd and curious little facts, and I have been really astounded at my own industry whilst reading my chapters on Inheritance and Selection. God knows when the book will ever be completed, for I find that I am very weak and on my best days cannot do more than one or one and a half hours' work. It is a good deal harder than writing about my dear climbing plants."

In this year he received the greatest honour which a scientific man can receive in this country—the Copley Medal of the Royal Society. It is presented at the Anniversary Meeting on St. Andrew's Day (Nov. 30), the medallist being usually present to receive it, but this the state of my father's health prevented. He wrote to Mr. Fox on this subject:—

"I was glad to see your hand-writing. The Copley, being open to all sciences and all the world, is reckoned a great honour; but excepting from several kind letters, such things make little difference to me. It shows, however, that

Natural Selection is making some progress in this country, and that pleases me. The subject, however, is safe in foreign lands."

To Sir J. D. Hooker, also, he wrote:—

"How kind you have been about this medal; indeed, I am blessed with many good friends, and I have received four or five notes which have warmed my heart. I often wonder that so old a worn-out dog as I am is not quite forgotten. Talking of medals, has Falconer had the Royal? he surely ought to have it, as ought John Lubbock. By the way, the latter tells me that some old members of the Royal are quite shocked at my having the Copley. Do you know who?"

He wrote to Mr. Huxley:—

"I must and will answer you, for it is a real pleasure for me to thank you cordially for your note. Such notes as this of yours, and a few others, are the real medal to me, and not the round bit of gold. These have given me a pleasure which will long endure; so believe in my cordial thanks for your note."

Sir Charles Lyell, writing to my father in November 1864 ('Life,' vol. ii. p. 384), speaks of the supposed malcontents as being afraid to crown anything so unorthodox as the 'Origin.' But he adds that if such were their feelings "they had the good sense to draw in their horns." It appears, however, from the same letter, that the proposal to give the Copley Medal to my father in the previous year failed owing to a similar want of courage—to Lyell's great indignation.

In the *Reader*, December 3, 1864, General Sabine's presidential address at the Anniversary Meeting is reported at some length. Special weight was laid on my father's work in Geology, Zoology, and Botany, but the 'Origin of Species' is praised chiefly as containing "a mass of observations," &c. It is curious that as in the case of his election to the French Institute, so in this case, he was honoured not for the great work of his life, but for his less important work in special lines. The paragraph in General Sabine's address which refers to the 'Origin of Species,' is as follows:—

"In his most recent work 'On the Origin of Species,' although opinions may be divided or undecided with respect to its merits in some respects, all will allow that it contains a mass of observations bearing upon the habits, structure, affinities, and distribution of animals, perhaps unrivalled for interest, minuteness, and patience of observation. Some amongst us may perhaps incline to accept the theory indicated by the title of this work, while others may perhaps incline to refuse, or at least to remit it to a future time, when increased knowledge shall afford stronger grounds for its ultimate acceptance or rejection. Speaking generally and collectively, we have expressly omitted it from the grounds of our award."

I believe I am right in saying that no little dissatisfaction at the President's manner of allusion to the 'Origin' was felt by some Fellows of the Society.

The presentation of the Copley Medal is of interest in another way, inasmuch as it led to Sir C. Lyell making, in his after-dinner speech, a "confession of faith as to the 'Origin.'" He wrote to my father ('Life,' vol. ii. p. 384), "I said I had been forced to give up my old faith without thoroughly seeing my way to a new one. But I think you would have been satisfied with the length I went."]

C. Darwin to T. H. Huxley.

Down, Oct. 3 [1864].

MY DEAR HUXLEY,—If I do not pour out my admiration of your article * on Kölliker, I shall explode. I never read

* "Criticisms on the Origin of Species," 'Nat. Hist. Review,' 1864. Republished in 'Lay Sermons,' 1870, p. 328. The work of Professor Kölliker referred to is 'Ueber die Darwin'sche Schöpfungstheorie' (Leipzig, 1864). Toward Professor Kölliker my father felt not only the respect due to so distinguished a naturalist (a sentiment well expressed in Professor Huxley's review), but he had also a personal regard for him, and often alluded with satisfaction to the visit which Professor Kölliker paid at Down.

anything better done. I had much wished his article answered, and indeed thought of doing so myself, so that I considered several points. You have hit on all, and on some in addition, and oh! by Jove, how well you have done it. As I read on and came to point after point on which I had thought, I could not help jeering and scoffing at myself, to see how infinitely better you had done it than I could have done. Well, if any one, who does not understand Natural Selection, will read this, he will be a blockhead if it is not as clear as daylight. Old Flourens * was hardly worth the powder and shot; but how capitally you bring in about the Academician, and your metaphor of the sea-sand is *inimitable.*

It is a marvel to me how you can resist becoming a regular reviewer. Well, I have exploded now, and it has done me a deal of good. . . .

[In the same article in the 'Natural History Review,' Mr. Huxley speaks of the book above alluded to by Flourens, the Secrétaire Perpétuel of the French Academy, as one of the two "most elaborate criticisms" of the 'Origin of Species' of the year. He quotes the following passage :—

"M. Darwin continue : 'Aucune distinction absolue n'a été et ne peut être établie entre les espèces et les variétés ! Je vous ai déjà dit que vous vous trompiez ; une distinction absolue sépare les variétés d'avec les espèces." Mr. Huxley remarks on this, "Being devoid of the blessings of an Academy in England, we are unaccustomed to see our ablest men treated in this way even by a Perpetual Secretary." After demonstrating M. Flourens' misapprehension of Natural Selection, Mr. Huxley says, "How one knows it all by heart, and with what relief one reads at p. 65, 'Je laisse M. Darwin.'"

On the same subject my father wrote to Mr. Wallace :—

"A great gun, Flourens, has written a little dull book

* 'Examen du livre de M. Darwin sur l'origine des espèces. Par P. Flourens.' 8vo. Paris, 1864.

against me, which pleases me much, for it is plain that our good work is spreading in France. He speaks of the 'engouement' about this book 'so full of empty and presumptuous thoughts.'" The passage here alluded to is as follows :—

"'Enfin l'ouvrage de M. Darwin a paru. On ne peut qu'être frappé du talent de l'auteur. Mais que d'idées obscures, que d'idées fausses! Quel jargon métaphysique jeté mal à propos dans l'histoire naturelle, qui tombe dans le galimatias dès qu'elle sort des idées claires, des idées justes. Quel langage prétentieux et vide! Quelles personnifications puériles et surannées! O lucidité! O solidité de l'esprit français, que devenez-vous?'"]

1865.

[This was again a time of much ill-health, but towards the close of the year he began to recover under the care of the late Dr. Bence-Jones, who dieted him severely, and as he expressed it, "half-starved him to death." He was able to work at 'Animals and Plants' until nearly the end of April, and from that time until December he did practically no work, with the exception of looking over the 'Origin of Species' for a second French edition. He wrote to Sir J. D. Hooker: —"I am, as it were, reading the 'Origin' for the first time, for I am correcting for a second French edition: and upon my life, my dear fellow, it is a very good book, but oh! my gracious, it is tough reading, and I wish it were done."*

The following letter refers to the Duke of Argyll's address to the Royal Society of Edinburgh, December 5th, 1864, in which he criticises the 'Origin of Species.' My father seems to have read the Duke's address as reported in the *Scotsman* of December 6th, 1865. In a letter to my father (Jan. 16,

* Towards the end of the year my father received the news of a new convert to his views, in the person of the distinguished American naturalist Lesquereux. He wrote to Sir J. D. Hooker: "I have had an enormous

1865, 'Life,' vol. ii. p. 385), Lyell wrote, "The address is a great step towards your views—far greater, I believe, than it seems when read merely with reference to criticisms and objections":]

C. Darwin to C. Lyell.

Down, January 22, 1865.

MY DEAR LYELL,—I thank you for your very interesting letter. I have the true English instinctive reverence for rank, and therefore liked to hear about the Princess Royal.* You ask what I think of the Duke's address, and I shall be glad to tell you. It seems to me *extremely* clever, like everything I have read of his; but I am not shaken—perhaps you will say that neither gods nor men could shake me. I demur to the Duke reiterating his objection that the brilliant plumage of the male humming-bird could not have been acquired through selection, at the same time entirely ignoring my discussion (p. 93, 3rd edition) on beautiful plumage being acquired through *sexual* selection. The Duke may think this insufficient, but that is another question. All analogy makes me quite disagree with the Duke that the difference in the beak, wing, and tail, are not of importance to the several species. In the only two species which I have watched, the difference in flight and in the use of the tail was conspicuously great.

The Duke, who knows my Orchid book so well, might have learnt a lesson of caution from it, with respect to his doctrine

letter from Leo Lesquereux (after doubts, I did not think it worth sending you) on Coal Flora. He wrote some excellent articles in 'Silliman' against 'Origin' views; but he says now, after repeated reading of the book, he is a convert!"

* "I had . . . an animated conversation on Darwinism with the Princess Royal, who is a worthy daughter of her father, in the reading of good books, and thinking of what she reads. She was very much *au fait* at the 'Origin,' and Huxley's book, the 'Antiquity,' &c."—Lyell's 'Life,' vol. ii. p. 385.

of differences for mere variety or beauty. It may be confidently said that no tribe of plants presents such grotesque and beautiful differences, which no one until lately, conjectured were of any use; but now in almost every case I have been able to show their important service. It should be remembered that with humming-birds or orchids, a modification in one part will cause correlated changes in other parts. I agree with what you say about beauty. I formerly thought a good deal on the subject, and was led quite to repudiate the doctrine of beauty being created for beauty's sake. I demur also to the Duke's expression of "new births." That may be a very good theory, but it is not mine, unless indeed he calls a bird born with a beak $\frac{1}{100}$th of an inch longer than usual "a new birth;" but this is not the sense in which the term would usually be understood. The more I work, the more I feel convinced that it is by the accumulation of such extremely slight variations that new species arise. I do not plead guilty to the Duke's charge, that I forget that natural selection means only the preservation of variations which independently arise.* I have expressed this in as strong language as I could use, but it would have been infinitely tedious had I on every occasion thus guarded myself. I will cry "peccavi" when I hear of the Duke or you attacking breeders for saying that man has made his improved shorthorns, or pouter pigeons, or bantams. And I could quote still stronger expressions used by agriculturists. Man does make his artificial breeds, for his selective power is of such importance relatively to that of the slight spontaneous variations. But no one will attack breeders for using such expressions, and the rising generation will not blame me.

Many thanks for your offer of sending me the 'Elements.'†

* "Strictly speaking, therefore, Mr. Darwin's theory is not a theory on the Origin of Species at all, but only a theory on the causes which lead to the relative success and failure of such new forms as may be born into the world."—*Scotsman*, Dec. 6, 1864.

† Sixth edition in one volume.

I hope to read it all, but unfortunately reading makes my head whiz more than anything else. I am able most days to work for two or three hours, and this makes all the difference in my happiness. I have resolved not to be tempted astray, and to publish nothing till my volume on Variation is completed. You gave me excellent advice about the footnotes in my Dog chapter, but their alteration gave me infinite trouble, and I often wished all the dogs, and I fear sometimes you yourself, in the nether regions.

We (dictator and writer) send our best love to Lady Lyell.

Yours affectionately,

CHARLES DARWIN.

P.S.—If ever you should speak with the Duke on the subject, please say how much interested I was with his address.

[In his autobiographical sketch, my father has remarked (p. 40) that owing to certain early memories he felt the honour of being elected to the Royal and Royal Medical Societies of Edinburgh "more than any similar honour." The following extract from a letter to Sir Joseph Hooker refers to his election to the former of these societies. The latter part of the extract refers to the Berlin Academy, to which he was elected in 1878 :—

"Here is a really curious thing, considering that Brewster is President and Balfour Secretary. I have been elected Honorary Member of the Royal Society of Edinburgh. And this leads me to a third question. Does the Berlin Academy of Sciences send their Proceedings to Honorary Members? I want to know, to ascertain whether I am a member; I suppose not, for I think it would have made some impression on me; yet I distinctly remember receiving some diploma signed by Ehrenberg. I have been so careless; I have lost several diplomas, and now I want to know what Societies I belong to, as I observe every [one] tacks their titles to their names in the catalogue of the Royal Soc."]

C. Darwin to C. Lyell.

Down, Feb. 21 [1865].

MY DEAR LYELL,—I have taken a long time to thank you very much for your present of the 'Elements.'

I am going through it all, reading what is new, and what I have forgotten, and this is a good deal.

I am simply astonished at the amount of labour, knowledge, and clear thought condensed in this work. The whole strikes me as something quite grand. I have been particularly interested by your account of Heer's work and your discussion on the Atlantic Continent. I am particularly delighted at the view which you take on this subject; for I have long thought Forbes did an ill service in so freely making continents.

I have also been very glad to read your argument on the denudation of the Weald, and your excellent *résumé* on the Purbeck Beds; and this is the point at which I have at present arrived in your book. I cannot say that I am quite convinced that there is no connection beyond that pointed out by you, between glacial action and the formation of lake basins; but you will not much value my opinion on this head, as I have already changed my mind some half-dozen times.

I want to make a suggestion to you. I found the weight of your volume intolerable, especially when lying down, so with great boldness cut it into two pieces, and took it out of its cover; now could not Murray without any other change add to his advertisement a line saying, "if bound in two volumes, one shilling or one shilling and sixpence extra." You thus might originate a change which would be a blessing to all weak-handed readers.

Believe me, my dear Lyell,

Yours most sincerely,

CHARLES DARWIN.

Originate a second *real blessing* and have the edges of the sheets cut like a bound book.*

C. Darwin to John Lubbock.

Down, June 11 [1865].

MY DEAR LUBBOCK,—The latter half of your book † has been read aloud to me, and the style is so clear and easy (we both think it perfection) that I am now beginning at the beginning. I cannot resist telling you how excellently well, in my opinion, you have done the very interesting chapter on savage life. Though you have necessarily only compiled the materials the general result is most original. But I ought to keep the term original for your last chapter, which has struck me as an admirable and profound discussion. It has quite delighted me, for now the public will see what kind of man you are, which I am proud to think I discovered a dozen years ago.

I do sincerely wish you all success in your election and in politics ; but after reading this last chapter, you must let me say: oh, dear! oh, dear! oh dear!

Yours affectionately,
CH. DARWIN.

P.S.—You pay me a superb compliment,‡ but I fear you

* This was a favourite reform of my father's. He wrote to the *Athenæum* on the subject, Feb. 5, 1867, pointing out that a book cut, even carefully, with a paper knife collects dust on its edges far more than a machine-cut book. He goes on to quote the case of a lady of his acquaintance who was in the habit of cutting books with her thumb, and finally appeals to the *Athenæum* to earn the gratitude of children "who have to cut through dry and pictureless books for the benefit of their elders." He tried to introduce the reform in the case of his own books, but found the conservatism of booksellers too strong for him. The presentation copies, however, of all his later books were sent out with the edges cut.

† 'Prehistoric Times,' 1865.

‡ 'Prehistoric Times,' p. 487, where the words, "the discoveries of a Newton or a Darwin," occur.

will be quizzed for it by some of your friends as too exaggerated.

[The following letter refers to Fritz Müller's book, 'Für Darwin,' which was afterwards translated, at my father's suggestion, by Mr. Dallas. It is of interest as being the first of the long series of letters which my father wrote to this distinguished naturalist. They never met, but the correspondence with Müller, which continued to the close of my father's life, was a source of very great pleasure to him. My impression is that of all his unseen friends Fritz Müller was the one for whom he had the strongest regard. Fritz Müller is the brother of another distinguished man, the late Hermann Müller, the author of 'Die Befruchtung der Blumen,' and of much other valuable work:]

C. Darwin to F. Müller.

Down, August 10 [1865].

MY DEAR SIR,—I have been for a long time so ill that I have only just finished hearing read aloud your work on species. And now you must permit me to thank you cordially for the great interest with which I have read it. You have done admirable service in the cause in which we both believe. Many of your arguments seem to me excellent, and many of your facts wonderful. Of the latter, nothing has surprised me so much as the two forms of males. I have lately investigated the cases of dimorphic plants, and I should much like to send you one or two of my papers if I knew how. I did send lately by post a paper on climbing plants, as an experiment to see whether it would reach you. One of the points which has struck me most in your paper is that on the differences in the air-breathing apparatus of the several forms. This subject appeared to me very important when I formerly considered the electric apparatus of fishes. Your observations on Classification and Embryology seem to me very good and original. They show what a wonderful field there is for enquiry on the development of crustacea, and nothing has convinced me so plainly what admirable results we shall arrive at in Natural History in the course of a few years. What a marvellous range of structure the crustacea present, and how well adapted they are for your enquiry! Until reading your book I knew nothing of the Rhizocephala; pray look at my account and figures of Anelasma, for it seems to me that this latter cirripede is a beautiful connecting link with the Rhizocephala.

If ever you have any opportunity, as you are so skilful a dissector, I much wish that you would look to the orifice at the base of the first pair of cirrhi in cirripedes, and at the curious organ in it, and discover what its nature is; I suppose I was quite in error, yet I cannot feel fully satisfied at Krohn's * observations. Also if you ever find any species of Scalpellum, pray look for complemental males; a German author has recently doubted my observations, for no reason except that the facts appeared to him so strange.

Permit me again to thank you cordially for the pleasure which I have derived from your work, and to express my sincere admiration for your valuable researches.

Believe me, dear Sir, with sincere respect,

Yours very faithfully,

CH. DARWIN.

P.S.—I do not know whether you care at all about plants, but if so, I should much like to send you my little work on the 'Fertilization of Orchids,' and I think I have a German copy.

Could you spare me a photograph of yourself? I should much like to possess one.

* See Vol. II. p. 345, Vol. III. p. 2.

C. Darwin to J. D. Hooker.

Down, Thursday, 27th [Sept. 1865].

MY DEAR HOOKER,—I had intended writing this morning to thank Mrs. Hooker most sincerely for her last and several notes about you, and now your own note in your hand has rejoiced me. To walk between five and six miles is splendid, with a little patience you must soon be well. I knew you had been very ill, but I hardly knew how ill, until yesterday, when Bentham (from the Cranworths *) called here, and I was able to see him for ten minutes. He told me also a little about the last days of your father; † I wish I had known your father better, my impression is confined to his remarkably cordial, courteous and frank bearing. I fully concur and understand what you say about the difference of feeling in the loss of a father and child. I do not think any one could love a father much more than I did mine, and I do not believe three or four days ever pass without my still thinking of him, but his death at eighty-four caused me nothing of that insufferable grief ‡ which the loss of poor dear Annie caused. And this seems to me perfectly natural, for one knows that for years previously

* Robert Rolfe, Lord Cranworth, and Lord Chancellor of England, lived at Holwood, near Down.

† Sir Wm. Hooker; b. 1785, d. 1865. He took charge of the Royal Gardens at Kew, in 1840, when they ceased to be the private gardens of the Royal Family. In doing so, he gave up his professorship at Glasgow—and with it half of his income. He founded the herbarium and library, and within ten years he succeeded in making the gardens the first in the world. It is, thus, not too much to say that the creation of the establishment at Kew is due to the abilities and self-devotion of Sir William Hooker. While, for the subsequent development of the gardens up to their present magnificent condition, the nation must thank Sir Joseph Hooker, in whom the same qualities are so conspicuous.

‡ I may quote here a passage from a letter of November 1863. It was written to a friend who had lost his child: "How well I remember your feeling, when we lost Annie. It was my greatest comfort that I had never spoken a harsh word to her. Your grief has made me shed a few tears over our poor darling; but believe me that these tears have lost that unutterable bitterness of former days."

that one's father's death is drawing slowly nearer and nearer, while the death of one's child is a sudden and dreadful wrench. What a wonderful deal you read; it is a horrid evil for me that I can read hardly anything, for it makes my head almost immediately begin to sing violently. My good womenkind read to me a great deal, but I dare not ask for much science, and am not sure that I could stand it. I enjoyed Tylor * *extremely*, and the first part of Lecky; † but I think the latter is often vague, and gives a false appearance of throwing light on his subject by such phrases as "spirit of the age," "spread of civilization," &c. I confine my reading to a quarter or half hour per day in skimming through the back volumes of the Annals and Magazines of Natural History, and find much that interests me. I miss my climbing plants very much, as I could observe them when very poorly.

I did not enjoy the 'Mill on the Floss' so much as you, but from what you say we will read it again. Do you know 'Silas Marner'? it is a charming little story; if you run short, and like to have it, we could send it by post. . . . We have almost finished the first volume of Palgrave,‡ and I like it much; but did you ever see a book so badly arranged? The frequency of the allusions to what will be told in the future are quite laughable. . . . By the way, I was very much pleased with the foot-note § about Wallace in Lubbock's last chapter. I had not heard that Huxley had backed up Lubbock about Parliament. . . . Did you see a sneer some time ago in the *Times* about how incomparably more interesting

* 'Researches into the Early History of Mankind,' by E. B. Tylor. 1865.

† 'The Rise of Rationalism in Europe,' by W. E. H. Lecky. 1865.

‡ William Gifford Palgrave's 'Travels in Arabia,' published in 1865.

§ The passage which seems to be referred to occurs in the text (p. 479) of 'Prehistoric Times.' It expresses admiration of Mr. Wallace's paper in the 'Anthropological Review' (May 1864), and speaks of the author's "characteristic unselfishness" in ascribing the theory of Natural Selection "unreservedly to Mr. Darwin."

politics were compared with science even to scientific men? Remember what Trollope says, in 'Can you Forgive her?' about getting into Parliament, as the highest earthly ambition. Jeffrey, in one of his letters, I remember, says that making an effective speech in Parliament is a far grander thing than writing the grandest history. All this seems to me a poor short-sighted view. I cannot tell you how it has rejoiced me once again seeing your handwriting—my best of old friends.

Yours affectionately,
CH. DARWIN.

[In October he wrote Sir J. D. Hooker:—

"Talking of the 'Origin,' a Yankee has called my attention to a paper attached to Dr. Wells' famous 'Essay on Dew,' which was read in 1813 to the Royal Soc., but not [then] printed, in which he applies most distinctly the principle of Natural Selection to the Races of Man. So poor old Patrick Matthew is not the first, and he cannot, or ought not, any longer to put on his title-pages, 'Discoverer of the principle of Natural Selection'!"]

*C. Darwin to F. W. Farrar.**

Down, Nov. 2 [1865?]

DEAR SIR,—As I have never studied the science of language, it may perhaps seem presumptuous, but I cannot resist the pleasure of telling you what interest and pleasure I have derived from hearing read aloud your volume.†

I formerly read Max Müller, and thought his theory (if it deserves to be called so) both obscure and weak; and now, after hearing what you say, I feel sure that this is the case, and that your cause will ultimately triumph. My indirect interest in your book has been increased from Mr. Hensleigh Wedgwood, whom you often quote, being my brother-in-law.

* Canon of Westminster.

† 'Chapters on Language,' 1865.

No one could dissent from my views on the modification of species with more courtesy than you do. But from the tenor of your mind I feel an entire and comfortable conviction (and which cannot possibly be disturbed) that if your studies led you to attend much to general questions in natural history you would come to the same conclusion that I have done.

Have you ever read Huxley's little book of Lectures? I would gladly send you a copy if you think you would read it.

Considering what Geology teaches us, the argument from the supposed immutability of specific types seems to me much the same as if, in a nation which had no old writings, some wise old savage was to say that his language had never changed; but my metaphor is too long to fill up.

Pray believe me, dear Sir, yours very sincerely obliged,

C. DARWIN.

1866.

[The year 1866 is given in my father's Diary in the following words:—

"Continued correcting chapters of 'Domestic Animals.'

March 1*st.*—Began on 4th edition of 'Origin' of 1250 copies (received for it £238), making 7500 copies altogether.

May 10*th.*—Finished 'Origin,' except revises, and began going over Chapter XIII. of 'Domestic Animals.'

Nov. 21*st.*—Finished 'Pangenesis.'

Dec. 21*st.*—Finished re-going over all chapters, and sent them to printers.

Dec. 22*nd.*—Began concluding chapter of book."

He was in London on two occasions for a week at a time, staying with his brother, and for a few days (May 29th–June 2nd) in Surrey; for the rest of the year he was at Down.

There seems to have been a gradual amendment in his health; thus he wrote to Mr. Wallace (January 1866):—"My health is so far improved that I am able to work one or two hours a day."

With respect to the 4th edition he wrote to Sir J. D. Hooker:—

"The new edition of the 'Origin' has caused me two great vexations. I forgot Bates's paper on variation,* but I remembered in time his mimetic work, and now, strange to say, I find I have forgotten your Arctic paper! I know how it arose; I indexed for my bigger work, and never expected that a new edition of the 'Origin' would be wanted.

"I cannot say how all this has vexed me. Everything which I have read during the last four years I find is quite washy in my mind." As far as I know, Mr. Bates's paper was not mentioned in the later editions of the 'Origin,' for what reason I cannot say.

In connection with his work on 'The Variation of Animals and Plants,' I give here extracts from three letters addressed to Mr. Huxley, which are of interest as giving some idea of the development of the theory of 'Pangenesis,' ultimately published in 1868 in the book in question:]

C. Darwin to T. H. Huxley.

Down, May 27, [1865?]

. . . I write now to ask a favour of you, a very great favour from one so hard worked as you are. It is to read thirty pages of MS., excellently copied out, and give me, not lengthened criticism, but your opinion whether I may venture to publish it. You may keep the MS. for a month or two. I would not ask this favour, but I *really* know no one else whose judgment on the subject would be final with me.

* This appears to refer to "Notes on South American Butterflies," Trans. Entomolog. Soc., vol. v. (N.S.).

The case stands thus: in my next book I shall publish long chapters on bud- and seminal-variation, on inheritance, reversion, effects of use and disuse, &c. I have also for many years speculated on the different forms of reproduction. Hencc it has come to be a passion with me to try to connect all such facts by some sort of hypothesis. The MS. which I wish to send you gives such a hypothesis; it is a very rash and crude hypothesis, yet it has been a considerable relief to my mind, and I can hang on it a good many groups of facts. I well know that a mere hypothesis, and this is nothing more, is of little value; but it is very useful to me as serving as a kind of summary for certain chapters. Now I earnestly wish for your verdict given briefly as, "Burn it"—or, which is the most favourable verdict I can hope for, "It does rudely connect together certain facts, and I do not think it will immediately pass out of my mind." If you can say this much, and you do not think it absolutely ridiculous, I shall publish it in my concluding chapter. Now will you grant me this favour? You must refuse if you are too much overworked.

I must say for myself that I am a hero to expose my hypothesis to the fiery ordeal of your criticism.

July 12, [1865?]

MY DEAR HUXLEY,—I thank you most sincerely for having so carefully considered my MS. It has been a real act of kindness. It would have annoyed me extremely to have re-published Buffon's views, which I did not know of, but I will get the book; and if I have strength I will also read Bonnet. I do not doubt your judgment is perfectly just, and I will try to persuade myself not to publish. The whole affair is much too speculative; yet I think some such view will have to be adopted, when I call to mind such facts as the inherited effects of use and disuse, &c. But I will try to be cautious. . . .

[1865?]

MY DEAR HUXLEY,—Forgive my writing in pencil, as I can do so lying down. I have read Buffon: whole pages are laughably like mine. It is surprising how candid it makes one to see one's views in another man's words. I am rather ashamed of the whole affair, but not converted to a no-belief. What a kindness you have done me with your "vulpine sharpness." Nevertheless, there is a fundamental distinction between Buffon's views and mine. He does not suppose that each cell or atom of tissue throws off a little bud; but he supposes that the sap or blood includes his "organic molecules," *which are ready formed*, fit to nourish each organ, and when this is fully formed, they collect to form buds and the sexual elements. It is all rubbish to speculate as I have done; yet, if I ever have strength to publish my next book, I fear I shall not resist "Pangenesis," but I assure you I will put it humbly enough. The ordinary course of development of beings, such as the Echinodermata, in which new organs are formed at quite remote spots from the analogous previous parts, seems to me extremely difficult to reconcile on any view except the free diffusion in the parent of the germs or gemmules of each separate new organ: and so in cases of alternate generation. But I will not scribble any more. Hearty thanks to you, you best of critics and most learned man. . . . . .

[The letters now take up the history of the year 1866.]

C. Darwin to A. R. Wallace.

Down, July 5 [1866].

MY DEAR WALLACE,—I have been much interested by your letter, which is as clear as daylight. I fully agree with all that you say on the advantages of H. Spencer's excellent expression of "the survival of the fittest."* This, however, had not occurred to me till reading your letter. It is, however, a great objection to this term that it cannot be used as a substantive governing a verb; and that this is a real objection I infer from H. Spencer continually using the words, natural selection. I formerly thought, probably in an exaggerated degree, that it was a great advantage to bring into connection natural and artificial selection; this indeed led me to use a term in common, and I still think it some advantage. I wish I had received your letter two months ago, for I would have worked in "the survival, &c.," often in the new edition of the 'Origin,' which is now almost printed off, and of which I will of course send you a copy. I will use the term in my next book on Domestic Animals, &c., from which, by the way, I plainly see that you expect *much* too much. The term Natural Selection has now been so largely used abroad and at home, that I doubt whether it could be given up, and with all its faults I should be sorry to see the attempt made. Whether it will be rejected must now depend "on the survival of the fittest." As in time the term must grow intelligible the objections to its use will grow weaker and weaker. I doubt whether the use of any term would have made the subject intelligible to some minds, clear as it is to others; for do we not see even to the present day Malthus on Population absurdly misunderstood? This reflection about Malthus has often comforted me when I have been vexed at the misstatement of my views. As for M. Janet,† he is a metaphysician, and such gentlemen are so acute that I think they often misunderstand common folk. Your criticism on the

* Extract from a letter of Mr. Wallace's, July 2, 1866: "The term 'survival of the fittest' is the plain expression of the fact; 'natural selection' is a metaphorical expression of it, and to a certain degree indirect and incorrect, since . . . Nature . . . does not so much select special varieties as exterminate the most unfavourable ones."

† This no doubt refers to Janet's 'Matérialisme Contemporaine.'

double sense * in which I have used Natural Selection is new to me and unanswerable; but my blunder has done no harm, for I do not believe that any one, excepting you, has ever observed it. Again, I agree that I have said too much about "favourable variations;" but I am inclined to think that you put the opposite side too strongly; if every part of every being varied, I do not think we should see the same end, or object, gained by such wonderfully diversified means.

I hope you are enjoying the country, and are in good health, and are working hard at your Malay Archipelago book, for I will always put this wish in every note I write to you, as some good people always put in a text. My health keeps much the same, or rather improves, and I am able to work some hours daily. With many thanks for your interesting letter,

Believe me, my dear Wallace, yours sincerely,

CH. DARWIN.

C. Darwin to J. D. Hooker.

Down, Aug. 30 [1866].

MY DEAR HOOKER,—I was very glad to get your note and the Notts. Newspaper. I have seldom been more pleased in my life than at hearing how successfully your lecture † went off. Mrs. H. Wedgwood sent us an account, saying that you read capitally, and were listened to with profound attention and great applause. She says, when your final

* "I find you use 'Natural Selection' in two senses; 1st, for the simple preservation of favourable and rejection of unfavourable variations, in which case it is equivalent to the 'survival of the fittest,'—and 2ndly, for the *effect* or *change* produced by this preservation."—Extract from Mr. Wallace's letter above quoted.

† At the Nottingham meeting of the British Association, Aug. 27, 1866. The subject of the lecture was 'Insular Floras.' See *Gardeners' Chronicle*, 1866.

allegory * began, "for a minute or two we were all mystified, and then came such bursts of applause from the audience. It was thoroughly enjoyed amid roars of laughter and noise, making a most brilliant conclusion."

I am rejoiced that you will publish your lecture, and felt sure that sooner or later it would come to this, indeed it would have been a sin if you had not done so. I am especially rejoiced as you give the arguments for occasional transport with such perfect fairness; these will now receive a fair share of attention, as coming from you, a professed botanist. Thanks also for Grove's address; as a whole it strikes me as very good and original, but I was disappointed in the part about Species; it dealt in such generalities that it would apply to any view or no view in particular. . . . .

And now farewell. I do most heartily rejoice at your success, and for Grove's sake at the brilliant success of the whole meeting.

Yours affectionately,

CHARLES DARWIN.

[The next letter is of interest, as giving the beginning of the connection which arose between my father and Professor Victor Carus. The translation referred to is the third German edition, made from the fourth English one. From this time forward Professor Carus continued to translate my father's books into German. The conscientious care with which this work was done was of material service, and I well remember the admiration (mingled with a tinge of vexation at his own shortcomings) with which my father used to receive the lists of oversights, &c., which Professor Carus dis-

* Sir Joseph Hooker allegorised the Oxford meeting of the British Association as the gathering of a tribe of savages who believed that the new moon was created afresh each month. The anger of the priests and medicine men at a certain heresy, according to which the new moon is but the offspring of the old one, is excellently given.

covered in the course of translation. The connection was not a mere business one, but was cemented by warm feelings of regard on both sides.]

C. Darwin to Victor Carus.

Down, November 10, 1866.

MY DEAR SIR,—I thank you for your extremely kind letter. I cannot express too strongly my satisfaction that you have undertaken the revision of the new edition, and I feel the honour which you have conferred on me. I fear that you will find the labour considerable, not only on account of the additions, but I suspect that Bronn's translation is very defective, at least I have heard complaints on this head from quite a large number of persons. It would be a great gratification to me to know that the translation was a really good one, such as I have no doubt you will produce. According to our English practice, you will be fully justified in entirely omitting Bronn's Appendix, and I shall be very glad of its omission. A new edition may be looked at as a new work. . . . . You could add anything of your own that you liked, and I should be much pleased. Should you make any additions or append notes, it appears to me that Nägeli, "Entstehung und Begriff," &c.,* would be worth noticing, as one of the most able pamphlets on the subject. I am, however, far from agreeing with him that the acquisition of certain characters which appear to be of no service to plants, offers any great difficulty, or affords a proof of some innate tendency in plants towards perfection. If you intend to notice this pamphlet, I should like to write hereafter a little more in detail on the subject.

. . . . I wish I had known, when writing my Historical

* 'Entstehung und Begriff der Naturhistorischen Art.' An Address given at a public meeting of the Royal Academy of Sciences at Munich, Mar. 28, 1865.

Sketch, that you had in 1853 published your views on the genealogical connection of past and present forms.

I suppose you have the sheets of the last English edition on which I marked with pencil all the chief additions, but many little corrections of style were not marked.

Pray believe that I feel sincerely grateful for the great service and honour which you do me by the present translation.

I remain, my dear Sir, yours very sincerely,

CHARLES DARWIN.

P.S.—I should be *very much* pleased to possess your photograph, and I send mine in case you should like to have a copy.

*C. Darwin to C. Nägeli.**

Down, June 12 [1866].

DEAR SIR,—I hope you will excuse the liberty which I take in writing to you. I have just read, though imperfectly, your 'Entstehung und Begriff,' and have been so greatly interested by it, that I have sent it to be translated, as I am a poor German scholar. I have just finished a new [4th] edition of my 'Origin,' which will be translated into German, and my object in writing to you is to say that if you should see this edition you would think that I had borrowed from you, without acknowledgment, two discussions on the beauty of flowers and fruit; but I assure you every word was printed off before I had opened your pamphlet. Should you like to possess a copy of either the German or English new edition, I should be proud to send one. I may add, with respect to the beauty of flowers, that I have already hinted the same views as you hold in my paper on Lythrum.

Many of your criticisms on my views are the best which I have met with, but I could answer some, at least to my own satisfaction; and I regret extremely that I had not read your

* Professor of Botany at Munich.

pamphlet before printing my new edition.* On one or two points, I think, you have a little misunderstood me, though I dare say I have not been cautious in expressing myself. The remark which has struck me most, is that on the position of the leaves not having been acquired through natural selection, from not being of any special importance to the plant. I well remember being formerly troubled by an analogous difficulty, namely, the position of the ovules, their anatropous condition, &c. It was owing to forgetfulness that I did not notice this difficulty in the 'Origin.' Although I can offer no explanation of such facts, and only hope to see that they may be explained, yet I hardly see how they support the doctrine of some law of necessary development, for it is not clear to me that a plant, with its leaves placed at some particular angle, or with its ovules in some particular position, thus stands higher than another plant. But I must apologise for troubling you with these remarks.

As I much wish to possess your photograph, I take the liberty of enclosing my own, and with sincere respect I remain, dear Sir,

Yours faithfully,
CH. DARWIN.

[I give a few extracts from letters of various dates showing my father's interest, alluded to in the last letter, in the problem of the arrangement of the leaves on the stems of plants. It may be added that Professor Schwendener of Berlin has successfully attacked the question in his 'Mechanische Theorie der Blattstellungen,' 1878.

To Dr. Falconer.

August 26 [1863].

"Do you remember telling me that I ought to study Phyllotaxy? well I have often wished you at the bottom of

* Nägeli's Essay is noticed in the 5th edition.

the sea; for I could not resist, and I muddled my brains with diagrams, &c., and specimens, and made out, as might have been expected, nothing. Those angles are a most wonderful problem and I wish I could see some one give a rational explanation of them."

To Dr. Asa Gray.

May 11 [1861].

"If you wish to save me from a miserable death, do tell me why the angles of $\frac{1}{2}$, $\frac{1}{3}$, $\frac{2}{5}$, $\frac{3}{8}$, &c., series occur, and no other angles. It is enough to drive the quietest man mad. Did you and some mathematician * publish some paper on the subject? Hooker says you did; where is it?

To Dr. Asa Gray.

[May 31, 1863?]

"I have been looking at Nägeli's work on this subject, and am astonished to see that the angle is not always the same in young shoots when the leaf-buds are first distinguishable, as in full-grown branches. This shows, I think, that there must be some potent cause for those angles which do occur: I dare say there is some explanation as simple as that for the angles of the Bees-cells."

My father also corresponded with Dr. Hubert Airy and was interested in his views on the subject, published in the Royal Soc. Proceedings, 1873, p. 176.

We now return to the year 1866. In November, when the prosecution of Governor Eyre was dividing England into two bitterly opposed parties, he wrote to Sir J. D. Hooker:—

* Probably my father was thinking of Chauncey Wright's work on Phyllotaxy, in Gould's 'Astronomical Journal,' No. 99, 1856, and in the 'Mathematical Monthly,' 1859. These papers are mentioned in the 'Letters of Chauncey Wright.' Mr. Wright corresponded with my father on the subject.

"You will shriek at me when you hear that I have just subscribed to the Jamaica Committee." *

On this subject I quote from a letter of my brother's :—

"With respect to Governor Eyre's conduct in Jamaica, he felt strongly that J. S. Mill was right in prosecuting him. I remember one evening, at my Uncle's, we were talking on the subject, and as I happened to think it was too strong a measure to prosecute Governor Eyre for murder, I made some foolish remark about the prosecutors spending the surplus of the fund in a dinner. My father turned on me almost with fury, and told me, if those were my feelings, I had better go back to Southampton; the inhabitants having given a dinner to Governor Eyre on his landing, but with which I had had nothing to do." The end of the incident, as told by my brother, is so characteristic of my father that I cannot resist giving it, though it has no bearing on the point at issue. "Next morning at 7 o'clock, or so, he came into my bedroom and sat on my bed, and said that he had not been able to sleep, from the thought that he had been so angry with me, and after a few more kind words he left me."

The same restless desire to correct a disagreeable or incorrect impression is well illustrated in a passage which I quote from some notes by Rev. J. Brodie Innes :—

"Allied to the extreme carefulness of observation was his most remarkable truthfulness in all matters. On one occasion, when a parish meeting had been held on some disputed point of no great importance, I was surprised by a visit from Mr. Darwin at night. He came to say that, thinking over the debate, though what he had said was quite accurate, he thought I might have drawn an erroneous conclusion, and he would not sleep till he had explained it. I believe that if on any day some certain fact had come to his knowledge which contradicted his most cherished theories, he would have placed the fact on record for publication before he slept."

* He subscribed £10.

This tallies with my father's habits, as described by himself. When a difficulty or an objection occurred to him, he thought it of paramount importance to make a note of it instantly, because he found hostile facts to be especially evanescent.

The same point is illustrated by the following incident, for which I am indebted to Mr. Romanes :—

" I have always remembered the following little incident as a good example of Mr. Darwin's extreme solicitude on the score of accuracy. One evening at Down there was a general conversation upon the difficulty of explaining the evolution of some of the distinctively human emotions, especially those appertaining to the recognition of beauty in natural scenery. I suggested a view of my own upon the subject, which, depending upon the principle of association, required the supposition that a long line of ancestors should have inhabited regions, the scenery of which is now regarded as beautiful. Just as I was about to observe that the chief difficulty attaching to my hypothesis arose from feelings of the sublime (seeing that these are associated with awe, and might therefore be expected not to be agreeable), Mr. Darwin anticipated the remark, by asking how the hypothesis was to meet the case of these feelings. In the conversation which followed, he said the occasion in his own life, when he was most affected by the emotions of the sublime was when he stood upon one of the summits of the Cordillera, and surveyed the magnificent prospect all around. It seemed, as he quaintly observed, as if his nerves had become fiddle-strings, and had all taken to rapidly vibrating. This remark was only made incidentally, and the conversation passed into some other branch. About an hour afterwards Mr. Darwin retired to rest, while I sat up in the smoking-room with one of his sons. We continued smoking and talking for several hours, when at about one o'clock in the morning the door gently opened and Mr. Darwin appeared, in his slippers and dressing-gown. As nearly as I can remember, the following are the words he used:—

"'Since I went to bed I have been thinking over our conversation in the drawing-room, and it has just occurred to me that I was wrong in telling you I felt most of the sublime when on the top of the Cordillera; I am quite sure that I felt it even more when in the forests of Brazil. I thought it best to come and tell you this at once in case I should be putting you wrong. I am sure now that I felt most sublime in the forests.'

"This was all he had come to say, and it was evident that he had come to do so, because he thought that the fact of his feeling 'most sublime in forests' was more in accordance with the hypothesis which we had been discussing, than the fact which he had previously stated. Now, as no one knew better than Mr. Darwin the difference between a speculation and a fact, I thought this little exhibition of scientific conscientiousness very noteworthy, where the only question concerned was of so highly speculative a character. I should not have been so much impressed if he had thought that by his temporary failure of memory he had put me on a wrong scent in any matter of fact, although even in such a case he is the only man I ever knew who would care to get out of bed at such a time of night in order to make the correction immediately, instead of waiting till next morning. But as the correction only had reference to a flimsy hypothesis, I certainly was very much impressed by this display of character."]

C. Darwin to J. D. Hooker.

Down, December 10 [1866].

. . . . I have now read the last No. of H. Spencer.* I do not know whether to think it better than the previous number, but it is wonderfully clever, and I dare say mostly true. I feel rather mean when I read him: I could bear, and rather enjoy

* 'Principles of Biology.'

feeling that he was twice as ingenious and clever as myself, but when I feel that he is about a dozen times my superior, even in the master art of wriggling, I feel aggrieved. If he had trained himself to observe more, even if at the expense, by the law of balancement, of some loss of thinking power, he would have been a wonderful man.

. . . . I am *heartily* glad you are taking up the Distribution of Plants in New Zealand, and suppose it will make part of your new book. Your view, as I understand it, that New Zealand subsided and formed two or more small islands, and then rose again, seems to me extremely probable. . . . . When I puzzled my brains about New Zealand, I remember I came to the conclusion, as indeed I state in the 'Origin,' that its flora, as well as that of other southern lands, had been tinctured by an Antarctic flora, which must have existed before the Glacial period. I concluded that New Zealand never could have been closely connected with Australia, though I supposed it had received some few Australian forms by occasional means of transport. Is there any reason to suppose that New Zealand could have been more closely connected with South Australia during the Glacial period, when the Eucalypti, &c., might have been driven further North? Apparently there remains only the line, which I think you suggested, of sunken islands from New Caledonia. Please remember that the Edwardsia was certainly drifted there by the sea.

I remember in old days speculating on the amount of life, *i.e.* of organic chemical change, at different periods. There seems to me one very difficult element in the problem, namely, the state of development of the organic beings at each period, for I presume that a Flora and Fauna of cellular cryptogamic plants, of Protozoa and Radiata would lead to much less chemical change than is now going on. But I have scribbled enough.

Yours affectionately,

CH. DARWIN.

[The following letter is in acknowledgment of Mr. Rivers' * reply to an earlier letter in which my father had asked for information on bud-variation. It may find a place here in illustration of the manner of my father's intercourse with those "whose avocations in life had to do with the rearing or use of living things" †—an intercourse which bore such good fruit in the 'Variation of Animals and Plants.' Mr. Dyer has some excellent remarks on the unexpected value thus placed on the apparently trivial facts disinterred from weekly journals, or amassed by correspondence. He adds: "Horticulturists who had . . . moulded plants almost at their will, at the impulse of taste or profit, were at once amazed and charmed to find that they had been doing scientific work, and helping to establish a great theory."]

C. Darwin to T. Rivers.

Down, December 28, [1866 ?]

MY DEAR SIR,—Permit me to thank you cordially for your most kind letter. For years I have read with interest every scrap which you have written in periodicals, and abstracted in MS. your book on Roses, and several times I thought I would write to you, but did not know whether you would think me too intrusive. I shall, indeed, be truly obliged for any information you can supply me on bud-variation or sports. When any extra difficult points occur to me in my present subject (which is a mass of difficulties), I will apply to you, but I will not be unreasonable. It is most true what you say that any one to study well the physiology of the life of plants, ought to have under his eye a multitude of plants. I have endeavoured to do what I can by comparing statements by many writers and observing what I could myself. Unfortunately few have

* The late Mr. Rivers was an eminent horticulturist and writer on horticulture.

† Mr. Dyer in 'Charles Darwin.' —*Nature Series*, 1882, p. 39.

observed like you have done. As you are so kind, I will mention one other point on which I am collecting facts; namely, the effect produced on the stock by the graft; thus, it is *said*, that the purple-leaved filbert affects the leaves of the common hazel on which it is grafted (I have just procured a plant to try), so variegated jessamine is *said* to affect its stock. I want these facts partly to throw light on the marvellous laburnums, Adami-trifacial oranges, &c. That laburnum case seems one of the strangest in physiology. I have now growing splendid, *fertile*, yellow laburnums (with a long raceme like the so-called Waterer's laburnum) from seed of yellow flowers on the *C. Adami*. To a man like myself, who is compelled to live a solitary life, and sees few persons, it is no slight satisfaction to hear that I have been able at all [to] interest by my books observers like yourself.

As I shall publish on my present subject, I presume, within a year, it will be of no use your sending me the shoots of peaches and nectarines which you so kindly offer; I have recorded your facts.

Permit me again to thank you cordially; I have not often in my life received a kinder letter.

My dear Sir, yours sincerely,

CH. DARWIN.

CHAPTER II.

THE PUBLICATION OF THE 'VARIATION OF ANIMALS AND PLANTS UNDER DOMESTICATION.'

JANUARY 1867, TO JUNE 1868.

AT the beginning of the year 1867 he was at work on the final chapter—"Concluding Remarks" of the 'Variation of Animals and Plants under Domestication,' which was begun after the rest of the MS. had been sent to the printers in the preceding December. With regard to the publication of the book he wrote to Mr. Murray, on January 3:—

"I cannot tell you how sorry I am to hear of the enormous size of my book.* I fear it can never pay. But I cannot shorten it now; nor, indeed, if I had foreseen its length, do I see which parts ought to have been omitted.

"If you are afraid to publish it, say so at once, I beg you, and I will consider your note as cancelled. If you think fit, get any one whose judgment you rely on, to look over some of the more legible chapters, namely, the Introduction, and on dogs and plants, the latter chapters being, in my opinion, the dullest in the book. . . . The list of chapters, and the inspection of a few here and there, would give a good judge

* On January 9 he wrote to Sir J. D. Hooker: "I have been these last few days vexed and annoyed to a foolish degree by hearing that my MS. on Dom. An. and Cult. Plants will make 2 vols., both bigger than the 'Origin.' The volumes will have to be full-sized octavo, so I have written to Murray to suggest details to be printed in small type. But I feel that the size is quite ludicrous in relation to the subject. I am ready to swear at myself and at every fool who writes a book."

a fair idea of the whole book. Pray do not publish blindly, as it would vex me all my life if I led you to heavy loss."

Mr. Murray referred the MS. to a literary friend, and, in spite of a somewhat adverse opinion, willingly agreed to publish the book. My father wrote :—

"Your note has been a great relief to me. I am rather alarmed about the verdict of your friend, as he is not a man of science. I think if you had sent the 'Origin' to an unscientific man, he would have utterly condemned it. I am, however, *very glad* that you have consulted any one on whom you can rely.

"I must add, that my 'Journal of Researches' was seen in MS. by an eminent semi-scientific man, and was pronounced unfit for publication."

The proofs were begun in March, and the last revise was finished on November 15th, and during this period the only intervals of rest were two visits of a week each at his brother Erasmus's house in Queen Anne Street. He notes in his Diary :—

"I began this book [in the] beginning of 1860 (and then had some MS.), but owing to interruptions from my illness, and illness of children ; from various editions of the 'Origin,' and Papers, especially Orchis book and Tendrils, I have spent four years and two months over it."

The edition of 'Animals and Plants' was of 1500 copies, and of these 1260 were sold at Mr. Murray's autumnal sale, but it was not published until January 30, 1868. A new edition of 1250 copies was printed in February of the same year.

In 1867 he received the distinction of being made a knight of the Prussian Order "Pour le Mérite." * He seems

* The Order "Pour le Mérite" was founded in 1740 by Frederick II. by the re-christening of an "Order of Generosity," founded in 1665. It was at one time strictly military, having been previously both civil and military, and in 1840 the Order was again opened to civilians. The order consists of thirty members of German extraction, but distinguished foreigners are admitted to a kind of extraordinary member-

not to have known how great the distinction was, for in June 1868 he wrote to Sir J. D. Hooker:—

"What a man you are for sympathy. I was made "Eques" some months ago, but did not think much about it. Now, by Jove, we all do; but you, in fact, have knighted me."

The letters may now take up the story.]

C. Darwin to J. D. Hooker.

Down, February 8 [1867].

MY DEAR HOOKER,—I am heartily glad that you have been offered the Presidentship of the British Association, for it is a great honour, and as you have so much work to do, I am equally glad that you have declined it. I feel, however, convinced that you would have succeeded very well; but if I fancy myself in such a position, it actually makes my blood run cold. I look back with amazement at the skill and taste with which the Duke of Argyll made a multitude of little speeches at Glasgow. By the way, I have not seen the Duke's book,* but I formerly thought that some of the articles which appeared in periodicals were very clever, but not very profound. One of these was reviewed in the *Saturday Review* † some years ago, and the fallacy of some main argument was admirably exposed, and I sent the article to you, and you agreed strongly with it. . . . There was the other day a rather good review of the Duke's book in the

ship. Faraday, Herschel, and Thomas Moore have belonged to it in this way. From the thirty members a chancellor is elected by the king (the first officer of this kind was Alexander v. Humboldt); and it is the duty of the chancellor to notify a vacancy in the Order to the remainder of the thirty, who then elect by vote the new member—but the king has technically the appointment in his own hands.

* 'The Reign of Law,' 1867.

† *Sat. Review*, Nov. 15, 1862, 'The *Edinburgh Review* on the Supernatural.' Written by my cousin, Mr. Henry Parker.

Spectator, and with a new explanation, either by the Duke or the reviewer (I could not make out which), of rudimentary organs, namely, that economy of labour and material was a great guiding principle with God (ignoring waste of seed and of young monsters, &c.), and that making a new plan for the structure of animals was thought, and thought was labour, and therefore God kept to a uniform plan, and left rudiments. This is no exaggeration. In short, God is a man, rather cleverer than us. . . . I am very much obliged for the *Nation* (returned by this post); it is *admirably* good. You say I always guess wrong, but I do not believe any one, except Asa Gray, could have done the thing so well. I would bet even, or three to two, that it is Asa Gray, though one or two passages staggered me.

I finish my book on 'Domestic Animals,' &c., by a single paragraph, answering, or rather throwing doubt, in so far as so little space permits, on Asa Gray's doctrine that each variation has been specially ordered or led along a beneficial line. It is foolish to touch such subjects, but there have been so many allusions to what I think about the part which God has played in the formation of organic beings,* that I thought it shabby to evade the question. . . . I have even received several letters on the subject. . . . I overlooked your sentence about Providence, and suppose I treated it as Buckland did his own theology, when his Bridgewater Treatise was read aloud to him for correction. . . .

* Prof. Judd allows me to quote from some notes which he has kindly given me:—"Lyell once told me that he had frequently been asked if Darwin was not one of the most unhappy of men, it being suggested that his outrage upon public opinion should have filled him with remorse." Sir Charles must have been able, I think, to give a conclusive answer on this point. Professor Judd continues:—

"I made a note of this and other conversations of Lyell's at the time. At the present time such statements must appear strange to any one who does not recollect the revolution in opinion which has taken place during the last 23 years [1882]."

[The following letter, from Mrs. Boole, is one of those referred to in the last letter to Sir J. D. Hooker:]

DEAR SIR,—Will you excuse my venturing to ask you a question, to which no one's answer but your own would be quite satisfactory?

Do you consider the holding of your theory of Natural Selection, in its fullest and most unreserved sense, to be inconsistent—I do not say with any particular scheme of theological doctrine—but with the following belief, namely:—

That knowledge is given to man by the direct inspiration of the Spirit of God.

That God is a personal and Infinitely good Being.

That the effect of the action of the Spirit of God on the brain of man is especially a moral effect.

And that each individual man has within certain limits a power of choice as to how far he will yield to his hereditary animal impulses, and how far he will rather follow the guidance of the Spirit, who is educating him into a power of resisting those impulses in obedience to moral motives?

The reason why I ask you is this: my own impression has always been, not only that your theory was perfectly *compatible* with the faith to which I have just tried to give expression, but that your books afforded me a clue which would guide me in applying that faith to the solution of certain complicated psychological problems which it was of practical importance to me as a mother to solve. I felt that you had supplied one of the missing links—not to say *the* missing link—between the facts of science and the promises of religion. Every year's experience tends to deepen in me that impression.

But I have lately read remarks on the probable bearing of your theory on religious and moral questions which have perplexed and pained me sorely. I know that the persons who make such remarks must be cleverer and wiser than myself. I cannot feel sure that they are mistaken, unless you will tell me so. And I think—I cannot know for certain—but I *think*—that if I were an author, I would rather that the humblest student of my works should apply to me directly in a difficulty, than that she should puzzle too long over adverse and probably mistaken or thoughtless criticisms.

At the same time I feel that you have a perfect right to refuse to answer such questions as I have asked you. Science must take her path, and Theology hers, and they will meet when and where and how God pleases, and you are in no sense responsible for it if the meeting-point should still be very far off. If I receive no answer to this letter I shall infer nothing from your silence, except that you felt I had no right to make such inquiries of a stranger.

[My father replied as follows:]

Down, December 14, 1866.

DEAR MADAM,—It would have gratified me much if I could have sent satisfactory answers to your questions, or, indeed, answers of any kind. But I cannot see how the belief that all organic beings, including man, have been genetically derived from some simple being, instead of having been separately created, bears on your difficulties. These, as it seems to me, can be answered only by widely different evidence from science, or by the so-called "inner consciousness." My opinion is not worth more than that of any other man who has thought on such subjects, and it would be folly in me to give it. I may, however, remark that it has always appeared to me more satisfactory to look at the immense amount of pain and suffering in this world as the inevitable result of the natural sequence of events, *i.e.* general laws, rather than from the direct intervention of God, though I am aware this is not logical with reference to an omniscient Deity. Your last question seems to resolve itself into the problem of free will and necessity, which has been found by most persons insoluble.

I sincerely wish that this note had not been as utterly valueless as it is. I would have sent full answers, though I have little time or strength to spare, had it been in my power.

I have the honour to remain, dear Madam,

Yours very faithfully,

CHARLES DARWIN.

P.S.—I am grieved that my views should incidentally have caused trouble to your mind, but I thank you for your judgment, and honour you for it, that theology and science should each run its own course, and that in the present case I am not responsible if their meeting-point should still be far off.

[The next letter discusses the 'Reign of Law,' referred to a few pages back:]

C. Darwin to C. Lyell.

Down, June 1 [1867].

. . . I am at present reading the Duke, and am *very much* interested by him; yet I cannot but think, clever as the whole is, that parts are weak, as when he doubts whether each curvature of the beak of humming-birds is of service to each species. He admits, perhaps too fully, that I have shown the use of each little ridge and shape of each petal in orchids, and how strange he does not extend the view to humming-birds. Still odder, it seems to me, all that he says on beauty, which I should have thought a nonentity, except in the mind of some sentient being. He might have as well said that love existed during the secondary or Palæozoic periods. I hope you are getting on with your book better than I am with mine, which kills me with the labour of correcting, and is intolerably dull, though I did not think so when I was writing it. A naturalist's life would be a happy one if he had only to observe, and never to write.

We shall be in London for a week in about a fortnight's time, and I shall enjoy having a breakfast talk with you.

Yours affectionately,

C. DARWIN.

[The following letter refers to the new and improved translation of the 'Origin,' undertaken by Professor Carus:]

C. Darwin to J. Victor Carus.

Down, February 17 [1867].

MY DEAR SIR,—I have read your preface with care. It seems to me that you have treated Bronn with complete respect and great delicacy, and that you have alluded to your own labour with much modesty. I do not think that any of Bronn's friends can complain of what you say and what you have done. For my own sake, I grieve that you have not added notes, as I am sure that I should have profited much by them; but as you have omitted Bronn's objections, I believe that you have acted with excellent judgment and fairness in leaving the text without comment to the independent verdict of the reader. I heartily congratulate you that the main part of your labour is over; it would have been to most men a very troublesome task, but you seem to have indomitable powers of work, judging from those two wonderful and most useful volumes on zoological literature* edited by you, and which I never open without surprise at their accuracy, and gratitude for their usefulness. I cannot sufficiently tell you how much I rejoice that you were persuaded to superintend the translation of the present edition of my book, for I have now the great satisfaction of knowing that the German public can judge fairly of its merits and demerits. . . . .

With my cordial and sincere thanks, believe me,

My dear Sir, yours very faithfully,

CH. DARWIN.

* 'Bibliotheca Zoologica,' 1861.

[The earliest letter which I have seen from my father to Professor Haeckel, was written in 1865, and from that time forward they corresponded (though not, I think, with any regularity) up to the end of my father's life. His friendship with Haeckel was not merely growth of correspondence, as was the case with some others, for instance, Fritz Müller. Haeckel paid more than one visit to Down, and these were thoroughly enjoyed by my father. The following letter will serve to show the strong feeling of regard which he entertained for his correspondent—a feeling which I have often heard him emphatically express, and which was warmly returned. The book referred to is Haeckel's 'Generelle Morphologie,' published in 1866, a copy of which my father received from the author in January 1867.

Dr. E. Krause * has given a good account of Professor Haeckel's services to the cause of Evolution. After speaking of the lukewarm reception which the 'Origin' met with in Germany on its first publication, he goes on to describe the first adherents of the new faith as more or less popular writers, not especially likely to advance its acceptance with the professorial or purely scientific world. And he claims for Haeckel that it was his advocacy of Evolution in his 'Radiolaria' (1862), and at the "Versammlung" of Naturalists at Stettin in 1863, that placed the Darwinian question for the first time publicly before the forum of German science, and his enthusiastic propagandism that chiefly contributed to its success.

Mr. Huxley, writing in 1869, paid a high tribute to Professor Haeckel as the Coryphæus of the Darwinian movement in Germany. Of his 'Generelle Morphologie,' "an attempt to work out the practical applications" of the doctrine of Evolution to their final results, he says that it has the "force and suggestiveness, and . . . systematising power of Oken without his extravagance." Professor Huxley also

* 'Charles Darwin und sein Verhältniss zu Deutschland,' 1885.

testifies to the value of Haeckel's 'Schöpfungs-Geschichte' as an exposition of the 'Generelle Morphologie' "for an educated public."

Again, in his 'Evolution in Biology,' * Mr. Huxley wrote: "Whatever hesitation may, not unfrequently, be felt by less daring minds, in following Haeckel in many of his speculations, his attempt to systematise the doctrine of Evolution, and to exhibit its influence as the central thought of modern biology, cannot fail to have a far-reaching influence on the progress of science."

In the following letter my father alludes to the somewhat fierce manner in which Professor Haeckel fought the battle of 'Darwinismus,' and on this subject Dr. Krause has some good remarks (p. 162). He asks whether much that happened in the heat of the conflict might not well have been otherwise, and adds that Haeckel himself is the last man to deny this. Nevertheless he thinks that even these things may have worked well for the cause of Evolution, inasmuch as Haeckel "concentrated on himself by his 'Ursprung des Menschen-Geschlechts,' his 'Generelle Morphologie,' and 'Schöpfungs-Geschichte,' all the hatred and bitterness which Evolution excited in certain quarters," so that, "in a surprisingly short time it became the fashion in Germany that Haeckel alone should be abused, while Darwin was held up as the ideal of forethought and moderation."]

C. Darwin to E. Haeckel.

Down, May 21, 1867.

DEAR HAECKEL.—Your letter of the 18th has given me great pleasure, for you have received what I said in the most kind and cordial manner. You have in part taken what I said much stronger than I had intended. It never occurred to me for a moment to doubt that your work, with the whole

* An article in the 'Encyclopædia Britannica,' 9th edit., reprinted in 'Science and Culture,' 1881, p. 298.

subject so admirably and clearly arranged, as well as fortified by so many new facts and arguments, would not advance our common object in the highest degree. All that I think is that you will excite anger, and that anger so completely blinds every one, that your arguments would have no chance of influencing those who are already opposed to our views. Moreover, I do not at all like that you, towards whom I feel so much friendship, should unnecessarily make enemies, and there is pain and vexation enough in the world without more being caused. But I repeat that I can feel no doubt that your work will greatly advance our subject, and I heartily wish it could be translated into English, for my own sake and that of others. With respect to what you say about my advancing too strongly objections against my own views, some of my English friends think that I have erred on this side; but truth compelled me to write what I did, and I am inclined to think it was good policy. The belief in the descent theory is slowly spreading in England,* even amongst those who can give no reason for their belief. No body of men were at first so much opposed to my views as the members of the London Entomological Society, but now I am assured that, with the exception of two or three old men, all the members concur with me to a certain extent. It has been a great disappointment to me that I have never received your long letter written to me from the Canary Islands. I am rejoiced to hear that your tour, which seems to have been a most interesting one, has done your health much good. I am working away at my new book, but make very slow progress, and the work tries my health, which is much the same as when you were here.

* In October 1867 he wrote to Mr. Wallace:—"Mr. Warrington has lately read an excellent and spirited abstract of the 'Origin' before the Victoria Institute, and as this is a most orthodox body, he has gained the name of the Devil's Advocate. The discussion which followed during three consecutive meetings is very rich from the nonsense talked. If you would care to see the number I could send it you."

Victor Carus is going to translate it, but whether it is worth translation, I am rather doubtful. I am very glad to hear that there is some chance of your visiting England this autumn, and all in this house will be delighted to see you here.

Believe me, my dear Haeckel,

Yours very sincerely,

CHARLES DARWIN.

C. Darwin to F. Müller.

Down, July 31 [1867].

MY DEAR SIR,—I received a week ago your letter of June 2, full as usual of valuable matter and specimens. It arrived at exactly the right time, for I was enabled to give a pretty full abstract of your observations on the plant's own pollen being poisonous. I have inserted this abstract in the proof-sheets in my chapter on sterility, and it forms the most striking part of my whole chapter.* I thank you very sincerely for the most interesting observations, which, however, I regret that you did not publish independently. I have been forced to abbreviate one or two parts more than I wished . . . Your letters always surprise me, from the number of points to which you attend. I wish I could make my letters of any interest to you, for I hardly ever see a naturalist, and live as retired a life as you in Brazil. With respect to mimetic plants, I remember Hooker many years ago saying he believed that there were many, but I agree with you that it would be most difficult to distinguish between mimetic resemblance and the effects of peculiar conditions. Who can say to which of these causes to attribute the several plants with heath-like foliage at the Cape of Good Hope? Is it not also a difficulty that quadrupeds appear to recognise plants more by their [scent] than their appearance?

* In 'The Variation of Animals and Plants.'

What I have just said reminds me to ask you a question. Sir J. Lubbock brought me the other day what appears to be a terrestrial Planaria (the first ever found in the northern hemisphere) and which was coloured exactly like our dark-coloured slugs. Now slugs are not devoured by birds, like the shell-bearing species, and this made me remember that I found the Brazilian Planariæ actually together with striped Vaginuli which I believe were similarly coloured. Can you throw any light on this? I wish to know, because I was puzzled some months ago how it would be possible to account for the bright colours of the Planariæ in reference to sexual selection. By the way, I suppose they are hermaphrodites.

Do not forget to aid me, if in your power, with answers to *any* of my questions on expression, for the subject interests me greatly. With cordial thanks for your never-failing kindness, believe me,

Yours very sincerely,
CHARLES DARWIN.

C. Darwin to C. Lyell.

Down, July 18 [1867].

MY DEAR LYELL,—Many thanks for your long letter. I am sorry to hear that you are in despair about your book; * I well know that feeling, but am now getting out of the lower depths. I shall be very much pleased, if you can make the least use of my present book, and do not care at all whether it is published before yours. Mine will appear towards the end of November of this year; you speak of yours as not coming out till November, 1868, which I hope may be an error. There is nothing about Man in my book which can interfere with you, so I will order all the completed clean sheets to be sent (and others as soon as ready) to you, but please observe you will not care for the first volume, which is a mere record

* The 2nd volume of the 10th edit. of the 'Principles.'

of the amount of variation; but I hope the second will be somewhat more interesting. Though I fear the whole must be dull.

I rejoice from my heart that you are going to speak out plainly about species. My book about Man, if published, will be short, and a large portion will be devoted to sexual selection, to which subject I alluded in the 'Origin' as bearing on Man. . . .

C. Darwin to C. Lyell.

Down, August 22 [1867].

MY DEAR LYELL,—I thank you cordially for your last two letters. The former one did me *real* good, for I had got so wearied with the subject that I could hardly bear to correct the proofs,* and you gave me fresh heart. I remember thinking that when you came to the Pigeon chapter you would pass it over as quite unreadable. Your last letter has interested me in very many ways, and I have been glad to hear about those horrid unbelieving Frenchmen. I have been particularly pleased that you have noticed Pangenesis. I do not know whether you ever had the feeling of having thought so much over a subject that you had lost all power of judging it. This is my case with Pangenesis (which is 26 or 27 years old), but I am inclined to think that if it be admitted as a probable hypothesis it will be a somewhat important step in Biology.

I cannot help still regretting that you have ever looked at the slips, for I hope to improve the whole a good deal. It is surprising to me, and delightful, that you should care in the least about the plants. Altogether you have given me one of the best cordials I ever had in my life, and I heartily thank you. I despatched this morning the French edition.† The

* The proofs of 'Animals and Plants,' which Lyell was then reading.

† Of the 'Origin.' It appears that my father was sending a copy of the French edition to Sir Charles. The introduction was by Mdlle. Royer, who translated the book.

introduction was a complete surprise to me, and I dare say has injured the book in France; nevertheless . . . it shows, I think, that the woman is uncommonly clever. Once again many thanks for the renewed courage with which I shall attack the horrid proof-sheets.

Yours affectionately,
CHARLES DARWIN.

P.S.—A Russian who is translating my new book into Russian has been here, and says you are immensely read in Russia, and many editions—how many I forget. Six editions of Buckle and four editions of the 'Origin.'

C. Darwin to Asa Gray.

Down, October 16 [1867].

MY DEAR GRAY,—I send by this post clean sheets of Vol. I. up to p. 336, and there are only 411 pages in this vol. I am *very* glad to hear that you are going to review my book; but if the *Nation* * is a newspaper I wish it were at the bottom of the sea, for I fear that you will thus be stopped reviewing me in a scientific journal. The first volume is all details, and you will not be able to read it; and you must remember that the chapters on plants are written for naturalists who are not botanists. The last chapter in Vol. I. is, however, I think, a curious compilation of facts; it is on bud-variation. In Vol. II. some of the chapters are more interesting; and I shall be very curious to hear your verdict on the chapter on close inter-breeding. The chapter on what I call Pangenesis will be called a mad dream, and I shall be pretty well satisfied if you think it a dream worth publishing; but at the bottom of my own mind I think it contains a great truth. I finish my book with a semi-theological paragraph, in which I quote and differ from you; what you will think of it, I know not. . . .

* The book was reviewed by Dr. Gray in the *Nation*, Mar. 19, 1868.

C. Darwin to J. D. Hooker.

Down, November 17 [1867].

MY DEAR HOOKER,—Congratulate me, for I have finished the last revise of the last sheet of my book. It has been an awful job : seven and a half months correcting the press : the book, from much small type, does not look big, but is really very big. I have had hard work to keep up to the mark, but during the last week only few revises came, so that I have rested and feel more myself. Hence, after our long mutual silence, I enjoy myself by writing a note to you, for the sake of exhaling, and hearing from you. On account of the index,* I do not suppose that you will receive your copy till the middle of next month. I shall be intensely anxious to hear what you think about Pangenesis ; though I can see how fearfully imperfect, even in mere conjectural conclusions, it is ; yet it has been an infinite satisfaction to me somehow to connect the various large groups of facts, which I have long considered, by an intelligible thread. I shall not be at all surprised if you attack it and me with unparalleled ferocity. It will be my endeavour to do as little as possible for some time, but [I] shall soon prepare a paper or two for the Linnean Society. In a short time we shall go to London for ten days, but the time is not yet fixed. Now I have told you a deal about myself, and do let me hear a good deal about your own past and future doings. Can you pay us a visit, early in December ? . . . . I have seen no one for an age, and heard no news.

. . . About my book I will give you a bit of advice. Skip the *whole* of Vol. I., except the last chapter (and that need only be skimmed) and skip largely in the 2nd volume ; and then you will say it is a very good book.

* The index was made by Mr. W. S. Dallas ; I have often heard my father express his admiration of this excellent piece of work.

1868.

['The Variation of Animals and Plants' was, as already mentioned, published on January 30, 1868, and on that day he sent a copy to Fritz Müller, and wrote to him :—

"I send by this post, by French packet, my new book, the publication of which has been much delayed. The greater part, as you will see, is not meant to be read; but I should very much like to hear what you think of 'Pangenesis,' though I fear it will appear to *every one* far too speculative."]

C. Darwin to J. D. Hooker.

February 3 [1868].

. . . I am very much pleased at what you say about my Introduction; after it was in type I was as near as possible cancelling the whole. I have been for some time in despair about my book, and if I try to read a few pages I feel fairly nauseated, but do not let this make you praise it; for I have made up my mind that it is not worth a fifth part of the enormous labour it has cost me. I assure you that all that is worth your doing (if you have time for so much) is glancing at Chapter VI., and reading parts of the later chapters. The facts on self-impotent plants seem to me curious, and I have worked out to my own satisfaction the good from crossing and evil from interbreeding. I did read Pangenesis the other evening, but even this, my beloved child, as I had fancied, quite disgusted me. The devil take the whole book; and yet now I am at work again as hard as I am able. It is really a great evil that from habit I have pleasure in hardly anything except Natural History, for nothing else makes me forget my ever-recurrent uncomfortable sensations. But I must not howl any more, and the critics may say what they like; I did my best, and man can do no more. What a splendid pursuit Natural History would be if it was all observing and no writing! . . . .

C. Darwin to J. D. Hooker.

Down, February 10 [1868].

My dear Hooker,—What is the good of having a friend, if one may not boast to him? I heard yesterday that Murray has sold in a week the whole edition of 1500 copies of my book, and the sale so pressing that he has agreed with Clowes to get another edition in fourteen days! This has done me a world of good, for I had got into a sort of dogged hatred of my book. And now there has appeared a review in the *Pall Mall* which has pleased me excessively, more perhaps than is reasonable. I am quite content, and do not care how much I may be pitched into. If by any chance you should hear who wrote the article in the *Pall Mall*, do please tell me; it is some one who writes capitally, and who knows the subject. I went to luncheon on Sunday, to Lubbock's, partly in hopes of seeing you, and, be hanged to you, you were not there.

Your cock-a-hoop friend,

C. D.

[Independently of the favourable tone of the able series of notices in the *Pall Mall Gazette* (Feb. 10, 15, 17, 1868), my father may well have been gratified by the following passages:—

"We must call attention to the rare and noble calmness with which he expounds his own views, undisturbed by the heats of polemical agitation which those views have excited, and persistently refusing to retort on his antagonists by ridicule, by indignation, or by contempt. Considering the amount of vituperation and insinuation which has come from the other side, this forbearance is supremely dignified."

And again in the third notice, Feb. 17:—

"Nowhere has the author a word that could wound the most sensitive self-love of an antagonist; nowhere does he, in text or note, expose the fallacies and mistakes of brother investigators . . . but while abstaining from impertinent censure, he is lavish in acknowledging the smallest debts he may owe; and his book will make many men happy."

I am indebted to Messrs. Smith & Elder for the information that these articles were written by Mr. G. H. Lewes.]

C. Darwin to J. D. Hooker.

Down, February 23 [1868].

MY DEAR HOOKER,—I have had almost as many letters to write of late as you can have, viz. from 8 to 10 per diem, chiefly getting up facts on sexual selection, therefore I have felt no inclination to write to you, and now I mean to write solely about my book for my own satisfaction, and not at all for yours. The first edition was 1500 copies, and now the second is printed off; sharp work. Did you look at the review in the *Athenæum*,* showing profound contempt of me? . . . It is a shame that he should have said that I have taken much from Pouchet, without acknowledgment; for I took literally nothing, there being nothing to take. There is a capital review in the *Gardeners' Chronicle*, which will sell the book if anything will.

* *Athenæum*, February 15, 1868. My father quoted Pouchet's assertion that "variation under domestication throws no light on the natural modification of species." The reviewer quotes the end of a passage in which my father declares that he can see no force in Pouchet's arguments, or rather assertions, and then goes on: "We are sadly mistaken if there are not clear proofs in the pages of the book before us that, on the contrary, Mr. Darwin has perceived, felt, and yielded to the force of the arguments or assertions of his French antagonist." The following may serve as samples of the rest of the review:—

"Henceforth the rhetoricians will have a better illustration of anti-climax than the mountain which brought forth a mouse, . . . in the discoverer of the origin of species, who tried to explain the variation of pigeons!

"A few summary words. On the 'Origin of Species' Mr. Darwin has nothing, and is never likely to have anything, to say; but on the vastly important subject of inheritance, the transmission of peculiarities once acquired through successive generations, this work is a valuable store-house of facts for curious students and practical breeders."

I don't quite see whether I or the writer is in a muddle about man *causing* variability. If a man drops a bit of iron into sulphuric acid he does not cause the affinities to come into play, yet he may be said to make sulphate of iron. I do not know how to avoid ambiguity.

After what the *Pall Mall Gazette* and the *Chronicle* have said, I do not care a d—.

I fear Pangenesis is stillborn; Bates says he has read it twice, and is not sure that he understands it. H. Spencer says the view is quite different from his (and this is a great relief to me, as I feared to be accused of plagiarism, but utterly failed to be sure what he meant, so thought it safest to give my view as almost the same as his), and he says he is not sure he understands it. . . . Am I not a poor devil? yet I took such pains, I must think that I expressed myself clearly. Old Sir H. Holland says he has read it twice, and thinks it very tough; but believes that sooner or later "some view akin to it" will be accepted.

You will think me very self-sufficient, when I declare that I feel *sure* if Pangenesis is now stillborn it will, thank God, at some future time reappear, begotten by some other father, and christened by some other name.

Have you ever met with any tangible and clear view of what takes place in generation, whether by seeds or buds, or how a long-lost character can possibly reappear; or how the male element can possibly affect the mother plant, or the mother animal, so that her future progeny are affected? Now all these points and many others are connected together, whether truly or falsely is another question, by Pangenesis. You see I die hard, and stick up for my poor child.

This letter is written for my own satisfaction, and not for yours. So bear it.

Yours affectionately,

CH. DARWIN.

*C. Darwin to A. Newton.**

Down, February 9 [1870].

DEAR NEWTON,—I suppose it would be universally held extremely wrong for a defendant to write to a Judge to express his satisfaction at a judgment in his favour; and yet I am going thus to act. I have just read what you have said in the 'Record' † about my pigeon chapters, and it has gratified me beyond measure. I have sometimes felt a little disappointed that the labour of so many years seemed to be almost thrown away, for you are the first man capable of forming a judgment (excepting partly Quatrefages), who seems to have thought anything of this part of my work. The amount of labour, correspondence, and care, which the subject cost me, is more than you could well suppose. I thought the article in the *Athenæum* was very unjust; but now I feel amply repaid, and I cordially thank you for your sympathy and too warm praise. What labour you have bestowed on your part of the 'Record'! I ought to be ashamed to speak of my amount of work. I thoroughly enjoyed the Sunday which you and the others spent here, and

I remain, dear Newton, yours very sincerely,

CH. DARWIN.

C. Darwin to A. R. Wallace.

Down, February 27 [1868].

MY DEAR WALLACE,—You cannot well imagine how much I have been pleased by what you say about 'Pangenesis.' None of my friends will speak out. . . . Hooker, as far as I understand him, which I hardly do at present, seems to think that the hypothesis is little more than saying that organisms have such and such potentialities. What you

* Prof. of Zoology at Cambridge.

† 'Zoological Record.' The volume for 1868, published Dec. 1869.

say exactly and fully expresses my feeling, viz. that it is a relief to have some feasible explanation of the various facts, which can be given up as soon as any better hypothesis is found. It has certainly been an immense relief to my mind; for I have been stumbling over the subject for years, dimly seeing that some relation existed between the various classes of facts. I now hear from H. Spencer that his views quoted in my foot-note refer to something quite distinct, as you seem to have perceived.

I shall be very glad to hear at some future day your criticisms on the "causes of variability." Indeed I feel sure that I am right about sterility and natural selection. . . . I do not quite understand your case, and we think that a word or two is misplaced. I wish some time you would consider the case under the following point of view:—If sterility is caused or accumulated through natural selection, then as every degree exists up to absolute barrenness, natural selection must have the power of increasing it. Now take two species, A and B, and assume that they are (by any means) half-sterile, *i.e.* produce half the full number of offspring. Now try and make (by natural selection) A and B absolutely sterile when crossed, and you will find how difficult it is. I grant, indeed it is certain, that the degree of sterility of the individuals A and B will vary, but any such extra-sterile individuals of, we will say A, if they should hereafter breed with other individuals of A, will bequeath no advantage to their progeny, by which these families will tend to increase in number over other families of A, which are not more sterile when crossed with B. But I do not know that I have made this any clearer than in the chapter in my book. It is a most difficult bit of reasoning, which I have gone over and over again on paper with diagrams.

. . . Hearty thanks for your letter. You have indeed pleased me, for I had given up the great god Pan as a still-born deity. I wish you could be induced to make it clear, with your admirable powers of elucidation, in one of the scientific journals. . . .

C. Darwin to J. D. Hooker.

Down, February 28 [1868].

MY DEAR HOOKER,—I have been deeply interested by your letter, and we had a good laugh over Huxley's remark, which was so deuced clever that you could not recollect it. I cannot quite follow your train of thought, for in the last page you admit all that I wish, having apparently denied all, or thought all mere words in the previous pages of your note; but it may be my muddle. I see clearly that any satisfaction which Pan may give will depend on the constitution of each man's mind. If you have arrived already at any similar conclusion, the whole will of course appear stale to you. I heard yesterday from Wallace, who says (excuse horrid vanity), "I can hardly tell you how much I admire the chapter on 'Pangenesis.' It is a *positive comfort* to me to have any feasible explanation of a difficulty that has always been haunting me, and I shall never be able to give it up till a better one supplies its place, and that I think hardly possible, &c." Now his foregoing [italicised] words express my sentiments exactly and fully: though perhaps I feel the relief extra strongly from having during many years vainly attempted to form some hypothesis. When you or Huxley say that a single cell of a plant, or the stump of an amputated limb, has the "potentiality" of reproducing the whole—or "diffuses an influence," these words give me no positive idea;—but, when it is said that the cells of a plant, or stump, include atoms derived from every other cell of the whole organism and capable of development, I gain a distinct idea. But this idea would not be worth a rush, if it applied to one case alone; but it seems to me to apply to all the forms of reproduction—inheritance—metamorphosis—to the abnormal transposition of organs—to the direct action of the male element on the mother plant, &c. Therefore I fully believe that each cell does *actually* throw off an atom or gemmule of its contents;—but whether or not, this hypothesis serves as a useful connecting link for various grand classes of physiological facts, which at present stand absolutely isolated.

I have touched on the doubtful point (alluded to by Huxley) how far atoms derived from the same cell may become developed into different structure accordingly as they are differently nourished; I advanced as illustrations galls and polypoid excrescences. . . .

It is a real pleasure to me to write to you on this subject, and I should be delighted if we can understand each other; but you must not let your good nature lead you on. Remember we always fight tooth and nail. We go to London on Tuesday, first for a week to Queen Anne Street, and afterwards to Miss Wedgwood's, in Regent's Park, and stay the whole month, which, as my gardener truly says, is a "terrible thing" for my experiments.

*C. Darwin to W. Ogle.**

Down, March 6 [1868].

DEAR SIR,—I thank you most sincerely for your letter, which is very interesting to me. I wish I had known of these views of Hippocrates before I had published, for they seem almost identical with mine—merely a change of terms—and an application of them to classes of facts necessarily unknown to the old philosopher. The whole case is a good illustration of how rarely anything is new.

. . . Hippocrates has taken the wind out of my sails, but I care very little about being forestalled. I advance the views

* Dr. William Ogle, now the Superintendent of Statistics to the Registrar-General.

merely as a provisional hypothesis, but with the secret expectation that sooner or later some such view will have to be admitted.

. . . I do not expect the reviewers will be so learned as you: otherwise, no doubt, I shall be accused of wilfully stealing Pangenesis from Hippocrates,—for this is the spirit some reviewers delight to show.

C. Darwin to Victor Carus.

Down, March 21 [1868].

. . . I am very much obliged to you for sending me so frankly your opinion on Pangenesis, and I am sorry it is unfavourable, but I cannot quite understand your remark on pangenesis, selection, and the struggle for life not being more methodical. I am not at all surprised at your unfavourable verdict; I know many, probably most, will come to the same conclusion. One English Review says it is much too complicated. . . . Some of my friends are enthusiastic on the hypothesis. . . . Sir C. Lyell says to every one, "You may not believe in 'Pangenesis,' but if you once understand it, you will never get it out of your mind." And with this criticism I am perfectly content. All cases of inheritance and reversion and development now appear to me under a new light. . . .

[An extract from a letter to Fritz Müller, though of later date (June), may be given here:—

"Your letter of April 22 has much interested me. I am delighted that you approve of my book, for I value your opinion more than that of almost any one. I have yet hopes that you will think well of Pangenesis. I feel sure that our minds are somewhat alike, and I find it a great relief to have some definite, though hypothetical view, when I reflect on the wonderful transformations of animals,—the re-growth of parts,—and especially the direct action of pollen on the mother-form, &c. It often appears to me almost certain that the characters of the parents are 'photographed' on the child, only by means of material atoms derived from each cell in both parents, and developed in the child."]

C. Darwin to Asa Gray.

Down, May 8 [1868].

My dear Gray,—I have been a most ungrateful and ungracious man not to have written to you an immense time ago to thank you heartily for the *Nation*, and for all your most kind aid in regard to the American edition [of 'Animals and Plants']. But I have been of late overwhelmed with letters, which I was forced to answer, and so put off writing to you. This morning I received the American edition (which looks capital), with your nice preface, for which hearty thanks. I hope to heaven that the book will succeed well enough to prevent you repenting of your aid. This arrival has put the finishing stroke to my conscience, which will endure its wrongs no longer.

. . . Your article in the *Nation* [Mar. 19] seems to me very good, and you give an excellent idea of Pangenesis—an infant cherished by few as yet, except his tender parent, but which will live a long life. There is parental presumption for you! You give a good slap at my concluding metaphor: * undoubtedly I ought to have brought in and contrasted natural and artificial selection; but it seemed so obvious to me that natural selection depended on contingencies even more

* A short abstract of the precipice metaphor is given at p. 307, vol. i. Dr. Gray's criticism on this point is as follows: "But in Mr. Darwin's parallel, to meet the case of nature according to his own view of it, not only the fragments of rock (answering to variation) should fall, but the edifice (answering to natural selection) should rise, irrespective of will or choice!" But my father's parallel demands that natural selection shall be the architect, not the edifice—the question of design only comes in with regard to the form of the building materials.

complex than those which must have determined the shape of each fragment at the base of my precipice. What I wanted to show was that, in reference to pre-ordainment, whatever holds good in the formation of a pouter pigeon holds good in the formation of a natural species of pigeon. I cannot see that this is false. If the right variations occurred, and no others, natural selection would be superfluous. A reviewer in an Edinburgh paper, who treats me with profound contempt, says on this subject that Professor Asa Gray could with the greatest ease smash me into little pieces.*

Believe me, my dear Gray,

Your ungrateful but sincere friend,

CHARLES DARWIN.

C. Darwin to G. Bentham.

Down, June 23, 1868.

MY DEAR MR. BENTHAM,—As your address † is somewhat of the nature of a verdict from a judge, I do not know whether it is proper for me to do so, but I must and will thank you for the pleasure which you have given me. I am delighted at what you say about my book. I got so tired of it, that for months together I thought myself a perfect fool for having given up so much time in collecting and observing little facts, but now I do not care if a score of common critics speak as contemptuously of the book as did the *Athenæum*. I feel justified in this, for I have so complete a reliance on your judgment that I feel certain that I should have bowed to your

* The *Daily Review*, April 27, 1868. My father has given rather a highly coloured version of the reviewer's remarks: "We doubt not that Professor Asa Gray . . . could show that natural selection . . . is simply an instrument in the hands of an omnipotent and omniscient creator." The reviewer goes on to say that the passage in question is a "very melancholy one," and that the theory is the "apotheosis of materialism."

† Presidential Address to the Linnean Society.

judgment had it been as unfavourable as it is the contrary. What you say about Pangenesis quite satisfies me, and is as much perhaps as any one is justified in saying. I have read your whole Address with the greatest interest. It must have cost you a vast amount of trouble. With cordial thanks, pray believe me,

Yours very sincerely,

CH. DARWIN.

P.S.—I fear that it is not likely that you have a superfluous copy of your Address; if you have, I should much like to send one to Fritz Müller in the interior of Brazil. By the way, let me add that I discussed bud-variation chiefly from a belief which is common to several persons, that all variability is related to sexual generation; I wished to show clearly that this was an error.

[The above series of letters may serve to show, to some extent, the reception which the new book received. Before passing on (in the next chapter) to the 'Descent of Man,' I give a letter referring to the translation of Fritz Müller's book, 'Für Darwin.' It was originally published in 1864, but the English translation, by Mr. Dallas, which bore the title suggested by Sir C. Lyell, of 'Facts and Arguments for Darwin,' did not appear until 1869:]

C. Darwin to F. Müller.

Down, March 16 [1868].

MY DEAR SIR,—Your brother, as you will have heard from him, felt so convinced that you would not object to a translation of 'Für Darwin,'* that I have ventured to arrange for a translation. Engelmann has very liberally offered me

* In a letter to Fritz Müller, my father wrote:—"I am vexed to see that on the title my name is more conspicuous than yours, which I especially objected to, and I cautioned the printers after seeing one proof."

clichés of the woodcuts for 22 thalers; Mr. Murray has agreed to bring out a translation (and he is our best publisher) on commission, for he would not undertake the work on his own risk; and I have agreed with Mr. W. S. Dallas (who has translated Von Siebold on Parthenogenesis, and many German works, and who writes very good English) to translate the book. He thinks (and he is a good judge) that it is important to have some few corrections or additions, in order to account for a translation appearing so lately [*i.e.* at such a long interval of time] after the original; so that I hope you will be able to send some. . . . .

[Two letters may be placed here, as bearing on the spread of Evolutionary ideas in France and Germany:]

C. Darwin to A. Gaudry.

Down, January 21 [1868].

DEAR SIR,—I thank you for your interesting essay on the influence of the Geological features of the country on the mind and habits of the Ancient Athenians,* and for your very obliging letter. I am delighted to hear that you intend to consider the relations of fossil animals in connection with their genealogy; it will afford you a fine field for the exercise of your extensive knowledge and powers of reasoning. Your belief will I suppose, at present, lower you in the estimation of your countrymen; but judging from the rapid spread in all parts of Europe, excepting France, of the belief in the common descent of allied species, I must think that this belief will before long become universal. How strange it is that the country which gave birth to Buffon, the elder Geoffroy, and especially to Lamarck, should now cling so pertinaciously to the belief that species are immutable creations.

* This appears to refer to M. Gaudry's paper translated in the 'Geol. Mag.,' 1868, p. 372.

My work on Variation, &c., under domestication, will appear in a French translation in a few months' time, and I will do myself the pleasure and honour of directing the publisher to send a copy to you to the same address as this letter.

With sincere respect, I remain, dear sir,

Yours very faithfully,

CHARLES DARWIN.

[The next letter is of especial interest, as showing how high a value my father placed on the support of the younger German naturalists :]

*C. Darwin to W. Preyer.**

March 31, 1868.

. . . . I am delighted to hear that you uphold the doctrine of the Modification of Species, and defend my views. The support which I receive from Germany is my chief ground for hoping that our views will ultimately prevail. To the present day I am continually abused or treated with contempt by writers of my own country ; but the younger naturalists are almost all on my side, and sooner or later the public must follow those who make the subject their special study. The abuse and contempt of ignorant writers hurts me very little. . . .

* Now Professor of Physiology at Jena.

CHAPTER III.

WORK ON 'MAN.

1864–1870.

[IN the autobiographical chapter (Vol. I. p. 93), my father gives the circumstances which led to his writing the 'Descent of Man.' He states that his collection of facts, begun in 1837 or 1838, was continued for many years without any definite idea of publishing on the subject. The following letter to Mr. Wallace shows that in the period of ill-health and depression about 1864 he despaired of ever being able to do so :]

C. Darwin to A. R. Wallace.

Down, [May ?] 28 [1864].

DEAR WALLACE,—I am so much better that I have just finished a paper for Linnean Society ;* but I am not yet at all strong, I felt much disinclination to write, and therefore you must forgive me for not having sooner thanked you for your paper on 'Man,'† received on the 11th. But first let me say that I have hardly ever in my life been more struck by any paper than that on 'Variation,' &c. &c., in the *Reader*.‡ I feel sure that such papers will do more for the spreading of

* On the three forms, &c., of Lythrum.

† 'Anthropological Review,' March 1864.

‡ *Reader*, Ap. 16, 1864. "On the Phenomena of Variation," &c. Abstract of a paper read before the Linnean Society, Mar. 17, 1864.

our views on the modification of species than any separate Treatises on the simple subject itself. It is really admirable; but you ought not in the Man paper to speak of the theory as mine; it is just as much yours as mine. One correspondent has already noticed to me your "high-minded" conduct on this head. But now for your Man paper, about which I should like to write more than I can. The great leading idea is quite new to me, viz. that during late ages, the mind will have been modified more than the body; yet I had got as far as to see with you, that the struggle between the races of man depended entirely on intellectual and *moral* qualities. The latter part of the paper I can designate only as grand and most eloquently done. I have shown your paper to two or three persons who have been here, and they have been equally struck with it. I am not sure that I go with you on all minor points: when reading Sir G. Grey's account of the constant battles of Australian savages, I remember thinking that natural selection would come in, and likewise with the Esquimaux, with whom the art of fishing and managing canoes is said to be hereditary. I rather differ on the rank, under a classificatory point of view, which you assign to man; I do not think any character simply in excess ought ever to be used for the higher divisions. Ants would not be separated from other hymenopterous insects, however high the instinct of the one, and however low the instincts of the other. With respect to the differences of race, a conjecture has occurred to me that much may be due to the correlation of complexion (and consequently hair) with constitution. Assume that a dusky individual best escaped miasma, and you will readily see what I mean. I persuaded the Director-General of the Medical Department of the Army to send printed forms to the surgeons of all regiments in tropical countries to ascertain this point, but I dare say I shall never get any returns. Secondly, I suspect that a sort of sexual selection has been the most powerful means of changing the races of man. I can show that the different races have a widely different standard of beauty. Among savages the most powerful men will have the pick of the women, and they will generally leave the most descendants. I have collected a few notes on man, but I do not suppose that I shall ever use them. Do you intend to follow out your views, and if so, would you like at some future time to have my few references and notes? I am sure I hardly know whether they are of any value, and they are at present in a state of chaos.

There is much more that I should like to write, but I have not strength.

Believe me, dear Wallace, yours very sincerely,

CH. DARWIN.

P.S.—Our aristocracy is handsomer (more hideous according to a Chinese or Negro) than the middle classes, from [having the] pick of the women; but oh, what a scheme is primogeniture for destroying natural selection! I fear my letter will be barely intelligible to you.

[In February 1867, when the manuscript of 'Animals and Plants' had been sent to Messrs. Clowes to be printed, and before the proofs began to come in, he had an interval of spare time, and began a "chapter on Man," but he soon found it growing under his hands, and determined to publish it separately as a "very small volume."

The work was interrupted by the necessity of correcting the proofs of 'Animals and Plants,' and by some botanical work, but was resumed with unremitting industry on the first available day in the following year. He could not rest, and he recognized with regret the gradual change in his mind that rendered continuous work more and more necessary to him as he grew older. This is expressed in a letter to Sir J. D. Hooker, June 17, 1868, which repeats to some extent what is given in the Autobiography:—

"I am glad you were at the 'Messiah,' it is the one thing that I should like to hear again, but I dare say I should find my soul too dried up to appreciate it as in old days; and then I should feel very flat, for it is a horrid bore to feel as I constantly do, that I am a withered leaf for every subject except Science. It sometimes makes me hate Science, though God knows I ought to be thankful for such a perennial interest, which makes me forget for some hours every day my accursed stomach."

The work on Man was interrupted by illness in the early summer of 1868, and he left home on July 16th for Freshwater, in the Isle of Wight, where he remained with his family until August 21st. Here he made the acquaintance of Mrs. Cameron. She received the whole family with open-hearted kindness and hospitality, and my father always retained a warm feeling of friendship for her. She made an excellent photograph of him, which was published with the inscription written by him: "I like this photograph very much better than any other which has been taken of me." Further interruption occurred in the autumn, so that continuous work on the 'Descent of Man' did not begin until 1869. The following letters give some idea of the earlier work in 1867:]

C. Darwin to A. R. Wallace.

Down, February 22, [1867?]

MY DEAR WALLACE,—I am hard at work on sexual selection, and am driven half mad by the number of collateral points which require investigation, such as the relative number of the two sexes, and especially on polygamy. Can you aid me with respect to birds which have strongly marked secondary sexual characters, such as birds of paradise, humming-birds, the Rupicola, or any other such cases? Many gallinaceous birds certainly are polygamous. I suppose that birds may be known not to be polygamous if they are seen during the whole breeding season to associate in pairs, or if the male incubates or aids in feeding the young. Will you have the kindness to turn this in your mind? But it is a shame to trouble you now that, as I am *heartily* glad to hear, you are at work on your Malayan travels. I am fearfully puzzled how far to extend your protective views with respect to the females in various classes. The more I work, the more important sexual selection apparently comes out.

Can butterflies be polygamous? *i.e.* will one male impregnate more than one female? Forgive me troubling you, and I dare say I shall have to ask forgiveness again. . . .

C. Darwin to A. R. Wallace.

Down, February 23 [1867].

DEAR WALLACE,—I much regretted that I was unable to call on you, but after Monday I was unable even to leave the house. On Monday evening I called on Bates, and put a difficulty before him, which he could not answer, and, as on some former similar occasion, his first suggestion was, "You had better ask Wallace." My difficulty is, why are caterpillars sometimes so beautifully and artistically coloured? Seeing that many are coloured to escape danger, I can hardly attribute their bright colour in other cases to mere physical conditions. Bates says the most gaudy caterpillar he ever saw in Amazonia (of a sphinx) was conspicuous at the distance of yards, from its black and red colours, whilst feeding on large green leaves. If any one objected to male butterflies having been made beautiful by sexual selection, and asked why should they not have been made beautiful as well as their caterpillars, what would you answer? I could not answer, but should maintain my ground. Will you think over this, and some time, either by letter or when we meet, tell me what you think? Also I want to know whether your *female* mimetic butterfly is more beautiful and brighter than the male. When next in London I must get you to show me your kingfishers. My health is a dreadful evil; I failed in half my engagements during this last visit to London.

Believe me, yours very sincerely,

C. DARWIN.

C. Darwin to A. R. Wallace.

Down, February 26 [1867].

MY DEAR WALLACE,—Bates was quite right; you are the man to apply to in a difficulty. I never heard anything more ingenious than your suggestion,* and I hope you may be able to prove it true. That is a splendid fact about the white moths; it warms one's very blood to see a theory thus almost proved to be true.† With respect to the beauty of male butterflies, I must as yet think that it is due to sexual selection. There is some evidence that dragon-flies are attracted by bright colours; but what leads me to the above belief, is so many male Orthoptera and Cicadas having musical instruments. This being the case, the analogy of birds makes me believe in sexual selection with respect to colour in insects. I wish I had strength and time to make some of the experiments suggested by you, but I thought butterflies would not pair in confinement. I am sure I have heard of some such difficulty. Many years ago I had a

* The suggestion that conspicuous caterpillars or perfect insects (e.g. white butterflies), which are distasteful to birds, are protected by being easily recognised and avoided. See Mr. Wallace's 'Natural Selection,' 2nd edit., p. 117.

† Mr. Jenner Weir's observations published in the Transactions of the Entomolog. Soc. (1869 and 1870) give strong support to the theory in question.

dragon-fly painted with gorgeous colours, but I never had an opportunity of fairly trying it.

The reason of my being so much interested just at present about sexual selection is, that I have almost resolved to publish a little essay on the origin of Mankind, and I still strongly think (though I failed to convince you, and this, to me, is the heaviest blow possible) that sexual selection has been the main agent in forming the races of man.

By the way, there is another subject which I shall introduce in my essay, namely, expression of countenance. Now, do you happen to know by any odd chance a very good-natured and acute observer in the Malay Archipelago, who you think would make a few easy observations for me on the expression of the Malays when excited by various emotions? For in this case I would send to such person a list of queries. I thank you for your most interesting letter, and remain,

Yours very sincerely,

CH. DARWIN.

C. Darwin to A. R. Wallace.

Down, March [1867].

MY DEAR WALLACE,—I thank you much for your two notes. The case of Julia Pastrana * is a splendid addition to my other cases of correlated teeth and hair, and I will add it in correcting the press of my present volume. Pray let me hear in the course of the summer if you get any evidence about the gaudy caterpillars. I should much like to give (or quote if published) this idea of yours, if in any way supported, as suggested by you. It will, however, be a long time hence, for I can see that sexual selection is growing into quite a large subject, which I shall introduce into my essay on Man, supposing that I ever publish it. I had

* A bearded woman having an irregular double set of teeth. See 'Animals and Plants,' vol. ii. p. 328.

intended giving a chapter on man, inasmuch as many call him (not *quite* truly) an eminently domesticated animal, but I found the subject too large for a chapter. Nor shall I be capable of treating the subject well, and my sole reason for taking it up is, that I am pretty well convinced that sexual selection has played an important part in the formation of races, and sexual selection has always been a subject which has interested me much. I have been very glad to see your impression from memory on the expression of Malays. I fully agree with you that the subject is in no way an important one; it is simply a "hobby-horse" with me, about twenty-seven years old; and *after* thinking that I would write an essay on Man, it flashed on me that I could work in some "supplemental remarks on expression." After the horrid, tedious, dull work of my present huge, and I fear unreadable, book ['The Variation of Animals and Plants'], I thought I would amuse myself with my hobby-horse. The subject is, I think, more curious and more amenable to scientific treatment than you seem willing to allow. I want, anyhow, to upset Sir C. Bell's view, given in his most interesting work, 'The Anatomy of Expression,' that certain muscles have been given to man solely that he may reveal to other men his feelings. I want to try and show how expressions have arisen. That is a good suggestion about newspapers, but my experience tells me that private applications are generally most fruitful. I will, however, see if I can get the queries inserted in some Indian paper. I do not know the names or addresses of any other papers.

. . . My two female amanuenses are busy with friends, and I fear this scrawl will give you much trouble to read. With many thanks,

Yours very sincerely,

CH. DARWIN.

[The following letter is worth giving, as an example

of his sources of information, and as showing what were the thoughts at this time occupying him :]

C. Darwin to F. Müller.

Down, June 3 [1868].

. . . Many thanks for all the curious facts about the unequal number of the sexes in Crustacea, but the more I investigate this subject the deeper I sink in doubt and difficulty. Thanks also for the confirmation of the rivalry of Cicadæ. I have often reflected with surprise on the diversity of the means for producing music with insects, and still more with birds. We thus get a high idea of the importance of song in the animal kingdom. Please to tell me where I can find any account of the auditory organs in the Orthoptera. Your facts are quite new to me. Scudder has described an insect in the Devonian strata, furnished with a stridulating apparatus. I believe he is to be trusted, and, if so, the apparatus is of astonishing antiquity. After reading Landois's paper I have been working at the stridulating organ in the Lamellicorn beetles, in expectation of finding it sexual; but I have only found it as yet in two cases, and in these it was equally developed in both sexes. I wish you would look at any of your common Lamellicorns, and take hold of both males and females, and observe whether they make the squeaking or grating noise equally. If they do not, you could, perhaps, send me a male and female in a light little box. How curious it is that there should be a special organ for an object apparently so unimportant as squeaking. Here is another point; have you any toucans? if so, ask any trustworthy hunter whether the beaks of the males, or of both sexes, are more brightly coloured during the breeding season than at other times of the year. . . . Heaven knows whether I shall ever live to make use of half the valuable facts which you have communicated to me! Your paper on *Balanus armatus*, translated by Mr. Dallas, has just appeared in our 'Annals and Magazine of Natural History,' and I have read it with the greatest interest. I never thought that I should live to hear of a hybrid Balanus! I am very glad that you have seen the cement tubes; they appear to me extremely curious, and, as far as I know, you are the first man who has verified my observations on this point.

With most cordial thanks for all your kindness, my dear Sir,

Yours very sincerely,
C. DARWIN.

C. Darwin to A. De Candolle.

Down, July 6, 1868.

MY DEAR SIR,—I return you my *sincere* thanks for your long letter, which I consider a great compliment, and which is quite full of most interesting facts and views. Your references and remarks will be of great use should a new edition of my book* be demanded, but this is hardly probable, for the whole edition was sold within the first week, and another large edition immediately reprinted, which I should think would supply the demand for ever. You ask me when I shall publish on the 'Variation of Species in a State of Nature.' I have had the MS. for another volume almost ready during several years, but I was so much fatigued by my last book that I determined to amuse myself by publishing a short essay on the 'Descent of Man.' I was partly led to do this by having been taunted that I concealed my views, but chiefly from the interest which I had long taken in the subject. Now this essay has branched out into some collateral subjects, and I suppose will take me more than a year to complete. I shall then begin on 'Species,' but my health makes me a very slow workman. I hope that you will excuse these details, which I have given to show

* 'Variation of Animals and Plants.'

that you will have plenty of time to publish your views first, which will be a great advantage to me. Of all the curious facts which you mention in your letter, I think that of the strong inheritance of the scalp-muscles has interested me most. I presume that you would not object to my giving this very curious case on your authority. As I believe all anatomists look at the scalp-muscles as a remnant of the *Panniculus carnosus* which is common to all the lower quadrupeds, I should look at the unusual development and inheritance of these muscles as probably a case of reversion. Your observation on so many remarkable men in noble families having been illegitimate is extremely curious; and should I ever meet any one capable of writing an essay on this subject I will mention your remarks as a good suggestion. Dr. Hooker has several times remarked to me that morals and politics would be very interesting if discussed like any branch of natural history, and this is nearly to the same effect with your remarks. . . .

C. Darwin to L. Agassiz.

Down, August 19, 1868.

DEAR SIR,—I thank you cordially for your very kind letter. I certainly thought that you had formed so low an opinion of my scientific work that it might have appeared indelicate in me to have asked for information from you, but it never occurred to me that my letter would have been shown to you. I have never for a moment doubted your kindness and generosity, and I hope you will not think it presumption in me to say, that when we met, many years ago, at the British Association at Southampton, I felt for you the warmest admiration.

Your information on the Amazonian fishes has interested me *extremely*, and tells me exactly what I wanted to know. I was aware, through notes given me by Dr. Günther, that many fishes differed sexually in colour and other characters, but I was particularly anxious to learn how far this was the case with those fishes in which the male, differently from what occurs with most birds, takes the largest share in the care of the ova and young. Your letter has not only interested me much, but has greatly gratified me in other respects, and I return you my sincere thanks for your kindness. Pray believe me, my dear Sir,

Yours very faithfully,
CHARLES DARWIN.

C. Darwin to J. D. Hooker.

Down, Sunday, August 23 [1868].

MY DEAR OLD FRIEND,—I have received your note. I can hardly say how pleased I have been at the success of your address,* and of the whole meeting. I have seen the *Times*, *Telegraph*, *Spectator*, and *Athenæum*, and have heard of other favourable newspapers, and have ordered a bundle. There is a "chorus of praise." The *Times* reported miserably, *i.e.* as far as errata were concerned; but I was very glad at the leader, for I thought the way you brought in the megalithic monuments most happy.† I particularly admired Tyndall's little speech.‡ . . . The *Spectator* pitches a little into you about Theology, in accordance with its usual spirit. . . .

Your great success has rejoiced my heart. I have just carefully read the whole address in the *Athenæum*; and though, as you know, I liked it very much when you read it to me, yet, as I was trying all the time to find fault, I missed to a certain extent the effect as a whole; and this now

* Sir Joseph Hooker was President of the British Association at the Norwich Meeting in 1868.

† The British Association was desirous of interesting the Government in certain modern cromlech builders, the Khasia race of East Bengal, in order that their megalithic monuments might be efficiently described.

‡ Professor Tyndall was President of Section A.

appears to me most striking and excellent. How you must rejoice at all your bothering labour and anxiety having had so grand an end. I must say a word about myself; never has such a eulogium been passed on me, and it makes me very proud. I cannot get over my *amazement* at what you say about my botanical work. By Jove, as far as my memory goes, you have strengthened instead of weakened some of the expressions. What is far more important than anything personal, is the conviction which I feel, that you will have immensely advanced the belief in the evolution of species. This will follow from the publicity of the occasion, your position, so responsible, as President, and your own high reputation. It will make a great step in public opinion, I feel sure, and I had not thought of this before. The *Athenæum* takes your snubbing * with the utmost mildness. I certainly do rejoice over the snubbing, and hope [the reviewer] will feel it a little. Whenever you have *spare* time to write again, tell me whether any astronomers † took your remarks in ill part; as they now stand they do not seem at all too harsh and presumptuous. Many of your sentences strike me as extremely felicitous and eloquent. That of Lyell's "underpinning," ‡ is capital. Tell me, was Lyell pleased? I am so glad that you remembered my old dedication.§ Was Wallace pleased?

* Sir Joseph Hooker made some reference to the review of 'Animals and Plants' in the *Athenæum* of Feb. 15, 1868.

† In discussing the astronomer's objection to Evolution, namely that our globe has not existed for a long enough period to give time for the assumed transmutation of living beings, Hooker challenged Whewell's dictum, that astronomy is the queen of sciences—the only perfect science.

‡ After a eulogium on Sir Charles Lyell's heroic renunciation of his old views in accepting Evolution, Sir J. D. Hooker continued, "Well may he be proud of a superstructure, raised on the foundations of an insecure doctrine, when he finds that he can underpin it and substitute a new foundation; and after all is finished, survey his edifice, not only more secure but more harmonious in its proportion than it was before."

§ The 'Naturalist's Voyage' was dedicated to Lyell.

How about photographs? Can you spare time for a line to our dear Mrs. Cameron?* She came to see us off, and loaded us with presents of photographs, and Erasmus called after her, "Mrs. Cameron, there are six people in this house all in love with you." When I paid her, she cried out, "Oh, what a lot of money!" and ran to boast to her husband.

I must not write any more, though I am in tremendous spirits at your brilliant success.

Yours ever affectionately,

C. DARWIN.

[In the *Athenæum* of November 29, 1868, appeared an article which was in fact a reply to Sir Joseph Hooker's remarks at Norwich. He seems to have consulted my father as to the wisdom of answering the article. My father wrote to him on December 1:—

"In my opinion Dr. Joseph Dalton Hooker need take no notice of the attack in the *Athenæum* in reference to Mr. Charles Darwin. What an ass the man is, to think he cuts one to the quick by giving one's Christian name in full. How transparently false is the statement that my sole groundwork is from pigeons, because I state I have worked them out more fully than other beings! He muddles together two books of Flourens."

The following letter refers to a paper† by Judge Caton, of which my father often spoke with admiration:]

C. Darwin to John D. Caton.

Down, September 18, 1868.

DEAR SIR,—I beg leave to thank you very sincerely for your kindness in sending me, through Mr. Walsh, your admirable paper on American Deer.

* See Vol. III. p. 92.

† 'Transactions of the Ottawa Academy of Natural Sciences,' 1868. By John D. Caton, late Chief Justice of Illinois.

It is quite full of most interesting observations, stated with the greatest clearness. I have seldom read a paper with more interest, for it abounds with facts of direct use for my work. Many of them consist of little points which hardly any one besides yourself has observed, or perceived the importance of recording. I would instance the age at which the horns are developed (a point on which I have lately been in vain searching for information), the rudiment of horns in the female elk, and especially the different nature of the plants devoured by the deer and elk, and several other points. With cordial thanks for the pleasure and instruction which you have afforded me, and with high respect for your power of observation, I beg leave to remain, dear Sir,

Yours faithfully and obliged,

CHARLES DARWIN.

[The following extract from a letter (Sept. 24, 1868) to the Marquis de Saporta, the eminent palæo-botanist, refers to the growth of Evolutionary views in France:—*

"As I have formerly read with great interest many of your papers on fossil plants, you may believe with what high satisfaction I hear that you are a believer in the gradual evolution of species. I had supposed that my book on the 'Origin of Species' had made very little impression in France, and therefore it delights me to hear a different statement from you. All the great authorities of the Institute seem firmly resolved to believe in the immutability of species, and this has always astonished me. . . . Almost the one exception, as far as I know, is M. Gaudry, and I think he will be soon one of the chief leaders in Zoological Palæontology in Europe; and now I am delighted to hear that in the sister department of Botany you take nearly the same view."]

* In 1868 he was pleased at being asked to authorise a French translation of his 'Naturalist's Voyage.'

C. Darwin to E. Haeckel.

Down, Nov. 19 [1868].

MY DEAR HAECKEL,—I must write to you again, for two reasons. Firstly, to thank you for your letter about your baby, which has quite charmed both me and my wife; I heartily congratulate you on its birth. I remember being surprised in my own case how soon the paternal instincts became developed, and in you they seem to be unusually strong, . . . I hope the large blue eyes and the principles of inheritance will make your child as good a naturalist as you are; but, judging from my own experience, you will be astonished to find how the whole mental disposition of your children changes with advancing years. A young child, and the same when nearly grown, sometimes differ almost as much as do a caterpillar and butterfly.

The second point is to congratulate you on the projected translation of your great work,* about which I heard from Huxley last Sunday. I am heartily glad of it, but how it has been brought about, I know not, for a friend who supported the proposed translation at Norwich, told me he thought there would be no chance of it. Huxley tells me that you consent to omit and shorten some parts, and I am confident that this is very wise. As I know your object is to instruct the public, you will assuredly thus get many more readers in England. Indeed, I believe that almost every book would be improved by condensation. I have been reading a good deal of your last book,† and the style is beautifully clear and easy to me; but why it should differ so much in this respect from your great work I cannot imagine. I have not yet read the first part, but began with the chapter on Lyell and myself, which you will easily believe

* 'Generelle Morphologie,' 1866. No English translation of this book has appeared.

† 'Die Natürliche Schöpfungs-Geschichte,' 1868. It was translated and published in 1876, under the title, 'The History of Creation.'

pleased me *very much*. I think Lyell, who was apparently much pleased by your sending him a copy, is also much gratified by this chapter.* Your chapters on the affinities and genealogy of the animal kingdom strike me as admirable and full of original thought. Your boldness, however, sometimes makes me tremble, but as Huxley remarked, some one must be bold enough to make a beginning in drawing up tables of descent. Although you fully admit the imperfection of the geological record, yet Huxley agreed with me in thinking that you are sometimes rather rash in venturing to say at what periods the several groups first appeared. I have this advantage over you, that I remember how wonderfully different any statement on this subject made 20 years ago, would have been to what would now be the case, and I expect the next 20 years will make quite as great a difference. Reflect on the monocotyledonous plant just discovered in the *primordial* formation in Sweden.

I repeat how glad I am at the prospect of the translation, for I fully believe that this work and all your works will have a great influence in the advancement of Science.

Believe me, my dear Häckel, your sincere friend,

CHARLES DARWIN.

[It was in November of this year that he sat for the bust by Mr. Woolner: he wrote :—

"I should have written long ago, but I have been pestered with stupid letters, and am undergoing the purgatory of sitting for hours to Woolner, who, however, is wonderfully pleasant, and lightens as much as man can, the penance ; as far as I can judge, it will make a fine bust."

If I may criticise the work of so eminent a sculptor as

* See Lyell's interesting letter to Haeckel. 'Life of Sir C. Lyell,' ii. p. 435.

Mr. Woolner, I should say that the point in which the bust fails somewhat as a portrait, is that it has a certain air, almost of pomposity, which seems to me foreign to my father's expression.]

1869.

[At the beginning of the year he was at work in preparing the fifth edition of the 'Origin.' This work was begun on the day after Christmas, 1868, and was continued for "forty-six days," as he notes in his diary, *i.e.* until February 10th, 1869. He then, February 11th, returned to Sexual Selection, and continued at this subject (excepting for ten days given up to Orchids, and a week in London), until June 10th, when he went with his family to North Wales, where he remained about seven weeks, returning to Down on July 31st.

Caerdeon, the house where he stayed, is built on the north shore of the beautiful Barmouth estuary, and is pleasantly placed in being close to wild hill country behind, as well as to the picturesque wooded "hummocks," between the steeper hills and the river. My father was ill and somewhat depressed throughout this visit, and I think felt saddened at being imprisoned by his want of strength, and unable even to reach the hills over which he had once wandered for days together.

He wrote from Caerdeon to Sir J. D. Hooker (June 22nd):—

"We have been here for ten days, how I wish it was possible for you to pay us a visit here; we have a beautiful house with a terraced garden, and a really magnificent view of Cader, right opposite. Old Cader is a grand fellow, and shows himself off superbly with every changing light. We remain here till the end of July, when the H. Wedgwoods have the house. I have been as yet in a very poor way; it seems as soon as the stimulus of mental work stops, my whole strength gives way. As yet I have hardly crawled half a mile from the house, and then have been fearfully fatigued. It is enough to make one wish oneself quiet in a comfortable tomb."

With regard to the fifth edition of the 'Origin,' he wrote to Mr. Wallace, January 22, 1869):—

"I have been interrupted in my regular work in preparing a new edition of the 'Origin,' which has cost me much labour, and which I hope I have considerably improved in two or three important points. I always thought individual differences more important than single variations, but now I have come to the conclusion that they are of paramount importance, and in this I believe I agree with you. Fleeming Jenkin's arguments have convinced me."

This somewhat obscure sentence was explained, February 2, in another letter to Mr. Wallace:—

"I must have expressed myself atrociously; I meant to say exactly the reverse of what you have understood. F. Jenkin argued in the 'North British Review' against single variations ever being perpetuated, and has convinced me, though not in quite so broad a manner as here put. I always thought individual differences more important; but I was blind and thought that single variations might be preserved much oftener than I now see is possible or probable. I mentioned this in my former note merely because I believed that you had come to a similar conclusion, and I like much to be in accord with you. I believe I was mainly deceived by single variations offering such simple illustrations, as when man selects."

The late Mr. Fleeming Jenkin's review, on the 'Origin of Species,' was published in the 'North British Review' for June 1867. It is not a little remarkable that the criticisms, which my father, as I believe, felt to be the most valuable ever made on his views should have come, not from a professed naturalist but from a Professor of Engineering.

It is impossible to give in a short compass an account of Fleeming Jenkin's argument. My father's copy of the paper (ripped out of the volume as usual, and tied with a bit of string) is annotated in pencil in many places. I may quote one passage opposite which my father has written "good sneers"—but it should be remembered that he used the word "sneer" in rather a special sense, not as necessarily implying a feeling of bitterness in the critic, but rather in the sense of "banter." Speaking of the 'true believer,' Fleeming Jenkin says, p. 293:—

"He can invent trains of ancestors of whose existence there is no evidence; he can marshal hosts of equally imaginary foes; he can call up continents, floods, and peculiar atmospheres; he can dry up oceans, split islands, and parcel out eternity at will; surely with these advantages he must be a dull fellow if he cannot scheme some series of animals and circumstances explaining our assumed difficulty quite naturally. Feeling the difficulty of dealing with adversaries who command so huge a domain of fancy, we will abandon these arguments, and trust to those which at least cannot be assailed by mere efforts of imagination."

In the fifth edition of the 'Origin,' my father altered a passage in the Historical Sketch (fourth edition, p. xviii). He thus practically gave up the difficult task of understanding whether or not Sir R. Owen claims to have discovered the principle of Natural Selection. Adding, "As far as the mere enunciation of the principle of Natural Selection is concerned, it is quite immaterial whether or not Professor Owen preceded me, for both of us . . . were long ago preceded by Dr. Wells and Mr. Matthew."

A somewhat severe critique on the fifth edition, by Mr. John Robertson, appeared in the *Athenæum*, August 14, 1869. The writer comments with some little bitterness on the success of the 'Origin:' "Attention is not acceptance. Many editions do not mean real success. The book has sold; the guess has been talked over; and the circulation and discussion sum up the significance of the editions." Mr. Robertson makes the true, but misleading statement: "Mr. Darwin prefaces his fifth English edition with an Essay, which he calls 'An Historical Sketch,' &c." As a matter of fact a Sketch appeared in the third edition in 1861.

Mr. Robertson goes on to say that the Sketch ought to be called a collection of extracts anticipatory or corroborative of the hypothesis of Natural Selection. "For no account is given of any hostile opinions. The fact is very significant. This historical sketch thus resembles the histories of the reign of Louis XVIII., published after the Restoration, from which the Republic and the Empire, Robespierre and Buonaparte were omitted."

The following letter to Prof. Victor Carus gives an idea of the character of the new edition of the 'Origin:']

C. Darwin to Victor Carus.

Down, May 4, 1869.

. . . I have gone very carefully through the whole, trying to make some parts clearer, and adding a few discussions and facts of some importance. The new edition is only two pages at the end longer than the old; though in one part nine pages in advance, for I have condensed several parts and omitted some passages. The translation I fear will cause you a great deal of trouble; the alterations took me six weeks, besides correcting the press; you ought to make a special agreement with M. Koch [the publisher]. Many of the corrections are only a few words, but they have been made from the evidence on various points appearing to have become a little stronger or weaker.

Thus I have been led to place somewhat more value on the definite and direct action of external conditions; to think the lapse of time, as measured by years, not quite so great as most geologists have thought; and to infer that single variations are of even less importance, in comparison with individual differences, than I formerly thought. I mention these points because I have been thus led to alter in many places *a few words;* and unless you go through the whole new edition, one part will not agree with another, which would be a great blemish. . . .

[The desire that his views might spread in France was always strong with my father, and he was therefore justly annoyed to find that in 1869 the publisher of the first French edition had brought out a third edition without consulting the author. He was accordingly glad to enter into an arrangement for a French translation of the fifth edition ; this was undertaken by M. Reinwald, with whom he continued to have pleasant relations as the publisher of many of his books into French.

He wrote to Sir J. D. Hooker:—

"I must enjoy myself and tell you about Mdlle. C. Royer, who translated the 'Origin' into French, and for whose second edition I took infinite trouble. She has now just brought out a third edition without informing me, so that all the corrections, &c., in the fourth and fifth English editions are lost. Besides her enormously long preface to the first edition, she has added a second preface abusing me like a pickpocket for Pangenesis, which of course has no relation to the 'Origin.' So I wrote to Paris ; and Reinwald agrees to bring out at once a new translation from the fifth English edition, in competition with her third edition. . . . This fact shows that "evolution of species" must at last be spreading in France."

With reference to the spread of Evolution among the orthodox, the following letter is of some interest. In March he received, from the author, a copy of a lecture by Rev. T. R. R. Stebbing, given before the Torquay Natural History Society, February 1, 1869, bearing the title "Darwinism." My father wrote to Mr. Stebbing :]

Down, March 3, 1869.

DEAR SIR,—I am very much obliged to you for your kindness in sending me your spirited and interesting lecture ;

if a layman had delivered the same address, he would have done good service in spreading what, as I hope and believe, is to a large extent the truth; but a clergyman in delivering such an address does, as it appears to me, much more good by his power to shake ignorant prejudices, and by setting, if I may be permitted to say so, an admirable example of liberality.

With sincere respect, I beg leave to remain,

Dear Sir, yours faithfully and obliged,

CHARLES DARWIN.

[The references to the subject of expression in the following letter are explained by the fact, that my father's original intention was to give his essay on this subject as a chapter in the 'Descent of Man,' which in its turn grew, as we have seen, out of a proposed chapter in 'Animals and Plants:']

C. Darwin to F. Müller.

Down, February 22, [1869?]

. . . Although you have aided me to so great an extent in many ways, I am going to beg for any information on two other subjects. I am preparing a discussion on "Sexual Selection," and I want much to know how low down in the animal scale sexual selection of a particular kind extends. Do you know of any lowly organised animals, in which the sexes are separated, and in which the male differs from the female in arms of offence, like the horns and tusks of male mammals, or in gaudy plumage and ornaments, as with birds and butterflies? I do not refer to secondary sexual characters, by which the male is able to discover the female, like the plumed antennæ of moths, or by which the male is enabled to seize the female, like the curious pincers described by you in some of the lower Crustaceans. But what I want to know is, how low in the scale sexual differences occur which require some degree of self-consciousness in the males, as weapons by which they fight for the female, or ornaments which attract the opposite sex. Any differences between males and females which follow different habits of life would have to be excluded. I think you will easily see what I wish to learn. *A priori*, it would never have been anticipated that insects would have been attracted by the beautiful colouring of the opposite sex, or by the sounds emitted by the various musical instruments of the male Orthoptera. I know no one so likely to answer this question as yourself, and should be grateful for any information, however small.

My second subject refers to expression of countenance, to which I have long attended, and on which I feel a keen interest; but to which, unfortunately, I did not attend, when I had the opportunity of observing various races of man. It has occurred to me that you might, without much trouble, make a *few* observations for me, in the course of some months, on Negroes, or possibly on native South Americans, though I care most about Negroes; accordingly I enclose some questions as a guide, and if you could answer me even one or two I should feel truly obliged. I am thinking of writing a little essay on the Origin of Mankind, as I have been taunted with concealing my opinions, and I should do this immediately after the completion of my present book. In this case I should add a chapter on the cause or meaning of expression. . . .

[The remaining letters of this year deal chiefly with the books, reviews, &c., which interested him.]

C. Darwin to H. Thiel.

Down, February 25, 1869.

DEAR SIR,—On my return home after a short absence, I found your very courteous note, and the pamphlet,* and I

* 'Ueber einige Formen der Landwirthschaftlichen Genossenschaften.' By Dr. H. Thiel, then of the Agricultural Station at Poppelsdorf.

hasten to thank you for both, and for the very honourable mention which you make of my name. You will readily believe how much interested I am in observing that you apply to moral and social questions analogous views to those which I have used in regard to the modification of species. It did not occur to me formerly that my views could be extended to such widely different, and most important, subjects. With much respect, I beg leave to remain, dear Sir,

Yours faithfully and obliged,

CHARLES DARWIN.

C. Darwin to T. H. Huxley.

Down, March 19 [1869].

MY DEAR HUXLEY,—Thanks for your 'Address.'* People complain of the unequal distribution of wealth, but it is a much greater shame and injustice that any one man should have the power to write so many brilliant essays as you have lately done. There is no one who writes like you. . . . If I were in your shoes, I should tremble for my life. I agree with all you say, except that I must think that you draw too great a distinction between the evolutionists and the uniformitarians.

I find that the few sentences which I have sent to press in the 'Origin' about the age of the world will do fairly well . . .

Ever yours,

C. DARWIN.

C. Darwin to A. R. Wallace.

Down, March 22 [1869].

MY DEAR WALLACE,—I have finished your book;† it seems to me excellent, and at the same time most pleasant to

* In his 'Anniversary Address' to the Geological Society, 1869, Mr. Huxley criticised Sir William Thomson's paper ('Trans. Geol. Soc. Glasgow,' vol. iii.) "On Geological Time."

† 'The Malay Archipelago,' &c. 1869.

read. That you ever returned alive is wonderful after all your risks from illness and sea voyages, especially that most interesting one to Waigiou and back. Of all the impressions which I have received from your book, the strongest is that your perseverance in the cause of science was heroic. Your descriptions of catching the splendid butterflies have made me quite envious, and at the same time have made me feel almost young again, so vividly have they brought before my mind old days when I collected, though I never made such captures as yours. Certainly collecting is the best sport in the world. I shall be astonished if your book has not a great success; and your splendid generalizations on Geographical Distribution, with which I am familiar from your papers, will be new to most of your readers. I think I enjoyed most the Timor case, as it is best demonstrated; but perhaps Celebes is really the most valuable. I should prefer looking at the whole Asiatic continent as having formerly been more African in its fauna, than admitting the former existence of a continent across the Indian Ocean. . . .

[The following letter refers to Mr. Wallace's article in the April number of the 'Quarterly Review,'* 1869, which to a large extent deals with the tenth edition of Sir Charles Lyell's 'Principles,' published in 1867 and 1868. The review contains a striking passage on Sir Charles Lyell's confession of evolutionary faith in the tenth edition of his 'Principles,' which is worth quoting: "The history of science hardly presents so striking an instance of youthfulness of mind in advanced life as is shown by this abandonment of opinions so long held and so powerfully advocated; and if we bear in mind the extreme caution, combined with the ardent love of truth

* My father wrote to Mr. Murray: "The article by Wallace is inimitably good, and it is a great triumph that such an article should appear in the 'Quarterly,' and will make the Bishop of Oxford and —— gnash their teeth."

which characterize every work which our author has produced, we shall be convinced that so great a change was not decided on without long and anxious deliberation, and that the views now adopted must indeed be supported by arguments of overwhelming force. If for no other reason than that Sir Charles Lyell in his tenth edition has adopted it, the theory of Mr. Darwin deserves an attentive and respectful consideration from every earnest seeker after truth."]

C. Darwin to A. R. Wallace.

Down, April 14, 1869.

MY DEAR WALLACE,—I have been wonderfully interested by your article, and I should think Lyell will be much gratified by it. I declare if I had been editor, and had the power of directing you, I should have selected for discussion the very points which you have chosen. I have often said to younger geologists (for I began in the year 1830) that they did not know what a revolution Lyell had effected; nevertheless, your extracts from Cuvier have quite astonished me. Though not able really to judge, I am inclined to put more confidence in Croll than you seem to do; but I have been much struck by many of your remarks on degradation. Thomson's views of the recent age of the world have been for some time one of my sorest troubles, and so I have been glad to read what you say. Your exposition of Natural Selection seems to me inimitably good; there never lived a better expounder than you. I was also much pleased at your discussing the difference between our views and Lamarck's. One sometimes sees the odious expression, "Justice to myself compels me to say," &c., but you are the only man I ever heard of who persistently does himself an injustice, and never demands justice. Indeed, you ought in the review to have alluded to your paper in the 'Linnean Journal,' and I feel sure all our friends will agree in this. But you cannot

I 2

"Burke" yourself, however much you may try, as may be seen in half the articles which appear. I was asked but the other day by a German professor for your paper, which I sent him. Altogether I look at your article as appearing in the 'Quarterly' as an immense triumph for our cause. I presume that your remarks on Man are those to which you alluded in your note. If you had not told me I should have thought that they had been added by some one else. As you expected, I differ grievously from you, and I am very sorry for it. I can see no necessity for calling in an additional and proximate cause in regard to man.* But the subject is too long for a letter. I have been particularly glad to read your discussion because I am now writing and thinking much about man.

I hope that your Malay book sells well; I was extremely pleased with the article in the 'Quarterly Journal of Science,' inasmuch as it is thoroughly appreciative of your work: alas! you will probably agree with what the writer says about the uses of the bamboo.

I hear that there is also a good article in the *Saturday Review*, but have heard nothing more about it. Believe me, my dear Wallace,

Yours ever sincerely,

CH. DARWIN.

C. Darwin to C. Lyell.

Down, May 4 [1869].

MY DEAR LYELL,—I have been applied to for some photo-

* Mr. Wallace points out that any one acquainted merely with the "unaided productions of nature," might reasonably doubt whether a dray-horse, for example, could have been developed by the power of man directing the "action of the laws of variation, multiplication, and survival, for his own purpose. We know, however, that this has been done, and we must therefore admit the possibility that in the development of the human race, a higher intelligence has guided the same laws for nobler ends."

graphs (carte de visite) to be copied to ornament the diplomas of honorary members of a new Society in Servia! Will you give me one for this purpose? I possess only a full-length one of you in my own album, and the face is too small, I think, to be copied.

I hope that you get on well with your work, and have satisfied yourself on the difficult point of glacier lakes. Thank heaven, I have finished correcting the new edition of the 'Origin,' and am at my old work of Sexual Selection.

Wallace's article struck me as *admirable;* how well he brought out the revolution which you effected some 30 years ago. I thought I had fully appreciated the revolution, but I was astounded at the extracts from Cuvier. What a good sketch of natural selection! but I was dreadfully disappointed about Man, it seems to me incredibly strange . . . ; and had I not known to the contrary, would have sworn it had been inserted by some other hand. But I believe that you will not agree quite in all this.

My dear Lyell, ever yours sincerely,

C. DARWIN.

C. Darwin to J. L. A. de Quatrefages.

Down, May 28 [1869 or 1870].

DEAR SIR,—I have received and read your volume,* and am much obliged for your present. The whole strikes me as a wonderfully clear and able discussion, and I was much interested by it to the last page. It is impossible that any account of my views could be fairer, or, as far as space permitted, fuller, than that which you have given. The way in which you repeatedly mention my name is most gratifying to me. When I had finished the second part, I thought that you had stated the case so favourably that you would make

* Essays reprinted from the 'Revue des Deux Mondes,' under the title 'Histoire Naturelle Générale,' &c., 1869.

more converts on my side than on your own side. On reading the subsequent parts I had to change my sanguine view. In these latter parts many of your strictures are severe enough, but all are given with perfect courtesy and fairness. I can truly say I would rather be criticised by you in this manner than praised by many others. I agree with some of your criticisms, but differ entirely from the remainder; but I will not trouble you with any remarks. I may, however, say, that you must have been deceived by the French translation, as you infer that I believe that the Parus and the Nuthatch (or Sitta) are related by direct filiation. I wished only to show, by an imaginary illustration, how either instincts or structures might first change. If you had seen *Canis Magellanicus* alive you would have perceived how foxlike its appearance is, or if you had heard its voice, I think that you would never have hazarded the idea that it was a domestic dog run wild; but this does not much concern me. It is curious how nationality influences opinion; a week hardly passes without my hearing of some naturalist in Germany who supports my views, and often puts an exaggerated value on my works; whilst in France I have not heard of a single zoologist, except M. Gaudry (and he only partially), who supports my views. But I must have a good many readers as my books are translated, and I must hope, notwithstanding your strictures, that I may influence some embryo naturalists in France.

You frequently speak of my good faith, and no compliment can be more delightful to me, but I may return you the compliment with interest, for every word which you write bears the stamp of your cordial love for the truth. Believe me, dear Sir, with sincere respect,

Yours very faithfully,

CHARLES DARWIN.

C. Darwin to T. H. Huxley:

Down, October 14, 1869.

MY DEAR HUXLEY,—I have been delighted to see your review of Häckel,* and as usual you pile honours high on my head. But I write now (*requiring no answer*) to groan a little over what you have said about rudimentary organs.† Many heretics will take advantage of what you have said. I cannot but think that the explanation given at p. 541 of the last edition of the 'Origin,' of the long retention of rudimentary organs and of their greater relative size during early life, is satisfactory. Their final and complete abortion seems to me a much greater difficulty. Do look in my 'Variations under Domestication,' vol. ii. p. 397, at what Pangenesis suggests on this head, though I did not dare to put it in the 'Origin.' The passage bears also a little on the struggle between the molecules or gemmules.‡ There is likewise a word or two indirectly bearing on this subject at pp. 394–395. It won't take you five minutes, so do look at these passages. I am very glad that you have been bold enough to give your idea about Natural Selection amongst the molecules, though I cannot quite follow you.

* A review of Haeckel's 'Schöpfungs-Geschichte.' The *Academy*, 1869. Reprinted in 'Critiques and Addresses,' p. 303.

† In discussing Teleology and Haeckel's "Dysteleology," Prof. Huxley says:—"Such cases as the existence of lateral rudiments of toes, in the foot of a horse, place us in a dilemma. For either these rudiments are of no use to the animals, in which case . . . they surely ought to have disappeared; or they are of some use to the animal, in which case they are of no use as arguments against Teleology."—'Critiques and Addresses,' p. 308.

‡ "It is a probable hypothesis, that what the world is to organisms in general, each organism is to the molecules of which it is composed. Multitudes of these having diverse tendencies, are competing with one another for opportunity to exist and multiply; and the organism, as a whole, is as much the product of the molecules which are victorious as the Fauna, or Flora, of a country is the product of the victorious organic beings in it."—'Critiques and Addresses,' p. 309.

1870.

[My father wrote in his Diary:—"The whole of this year [1870] at work on the 'Descent of Man.' . . . Went to Press August 30, 1870."

The letters are again of miscellaneous interest, dealing, not only with his work, but also serving to indicate the course of his reading.]

C. Darwin to E. Ray Lankester.

Down, March 15 [1870].

MY DEAR SIR,—I do not know whether you will consider me a very troublesome man, but I have just finished your book,* and cannot resist telling you how the whole has much interested me. No doubt, as you say, there must be much speculation on such a subject, and certain results cannot be reached; but all your views are highly suggestive, and to my mind that is high praise. I have been all the more interested, as I am now writing on closely allied though not quite identical points. I was pleased to see you refer to my much despised child, 'Pangenesis,' who I think will some day, under some better nurse, turn out a fine stripling. It has also pleased me to see how thoroughly you appreciate (and I do not think that this is general with the men of science) H. Spencer; I suspect that hereafter he will be looked at as by far the greatest living philosopher in England; perhaps equal to any that have lived. But I have no business to trouble you with my notions. With sincere thanks for the interest which your work has given me,

I remain, yours very faithfully,

CH. DARWIN.

[The next letter refers to Mr. Wallace's 'Natural Selec-

* 'Comparative Longevity.'

tion' (1870), a collection of essays reprinted with certain alterations of which a list is given in the volume:]

C. Darwin to A. R. Wallace.

Down, April 20 [1870].

MY DEAR WALLACE,—I have just received your book, and read the preface. There never has been passed on me, or indeed on any one, a higher eulogium than yours. I wish that I fully deserved it. Your modesty and candour are very far from new to me. I hope it is a satisfaction to you to reflect—and very few things in my life have been more satisfactory to me—that we have never felt any jealousy towards each other, though in one sense rivals. I believe that I can say this of myself with truth, and I am absolutely sure that it is true of you.

You have been a good Christian to give a list of your additions, for I want much to read them, and I should hardly have had time just at present to have gone through all your articles. Of course I shall immediately read those that are new or greatly altered, and I will endeavour to be as honest as can reasonably be expected. Your book looks remarkably well got up.

Believe me, my dear Wallace, to remain,

Yours very cordially,

CH. DARWIN.

[Here follow one or two letters indicating the progress of the 'Descent of Man;' the woodcuts referred to were being prepared for that work:]

*C. Darwin to A. Günther.**

March 23, [1870?]

DEAR GÜNTHER,—As I do not know Mr. Ford's address, will you hand him this note, which is written solely to express

* Dr. Günther, Keeper of Zoology in the British Museum.

my unbounded admiration of the woodcuts. I fairly gloat over them. The only evil is that they will make all the other woodcuts look very poor! They are all excellent, and for the feathers I declare I think it the most wonderful woodcut I ever saw; I cannot help touching it to make sure that it is smooth. How I wish to see the two other, and even more important, ones of the feathers, and the four [of] reptiles, &c. Once again accept my very sincere thanks for all your kindness. I am greatly indebted to Mr. Ford. Engravings have always hitherto been my greatest misery, and now they are a real pleasure to me.

Yours very sincerely,

CH. DARWIN.

P.S.—I thought I should have been in press by this time, but my subject has branched off into sub-branches, which have cost me infinite time, and heaven knows when I shall have all my MS. ready; but I am never idle.

C. Darwin to A. Günther.

May 15 [1870].

MY DEAR DR. GÜNTHER,—Sincere thanks. Your answers are wonderfully clear and complete. I have some analogous questions on reptiles, &c., which I will send in a few days, and then I think I shall cause no more trouble. I will get the books you refer me to. The case of the Solenostoma * is magnificent, so exactly analogous to that of those birds in which the female is the more gay, but ten times better for me, as she is the incubator. As I crawl on with the successive

* In most of the Lophobranchii the male has a marsupial sack in which the eggs are hatched, and in these species the male is slightly brighter coloured than the female. But in Solenostoma the female is the hatcher, and is also the more brightly coloured.—'Descent of Man,' ii. 21.

classes I am astonished to find how similar the rules are about the nuptial or "wedding dress" of all animals. The subject has begun to interest me in an extraordinary degree; but I must try not to fall into my common error of being too speculative. But a drunkard might as well say he would drink a little and not too much! My essay, as far as fishes, batrachians and reptiles are concerned, will be in fact yours, only written by me. With hearty thanks,

Yours very sincerely,

CH. DARWIN.

[The following letter is of interest, as showing the excessive care and pains which my father took in forming his opinion on a difficult point:]

C. Darwin to A. R. Wallace.

Down, September 23 [undated].

MY DEAR WALLACE,—I am very much obliged for all your trouble in writing me your long letter, which I will keep by me and ponder over. To answer it would require at least 200 folio pages! If you could see how often I have re-written some pages you would know how anxious I am to arrive as near as I can to the truth. I lay great stress on what I know takes place under domestication; I think we start with different fundamental notions on inheritance. I find it is most difficult, but not I think impossible, to see how, for instance, a few red feathers appearing on the head of a male bird, and which *are at first transmitted to both sexes*, could come to be transmitted to males alone. It is not enough that females should be produced from the males with red feathers, which should be destitute of red feathers; but these females must have a *latent tendency* to produce such feathers, otherwise they would cause deterioration in the red head-feathers of their male offspring. Such latent tendency would be shown by their producing the red feathers when old, or diseased in their ovaria. But I have no difficulty in making the whole head red if the few red feathers in the male from the first tended to be sexually transmitted. I am quite willing to admit that the female may have been modified, either at the same time or subsequently, for protection by the accumulation of variations limited in their transmission to the female sex. I owe to your writings the consideration of this latter point. But I cannot yet persuade myself that females *alone* have often been modified for protection. Should you grudge the trouble briefly to tell me, whether you believe that the plainer head and less bright colours of ♀ chaffinch,* the less red on the head and less clean colours of ♀ goldfinch, the much less red on the breast of ♀ bullfinch, the paler crest of golden-crested wren, &c., have been acquired by them for protection. I cannot think so, any more than I can that the considerable differences between ♀ and ♂ house sparrow, or much greater brightness of ♂ *Parus cæruleus* (both of which build under cover) than of ♀ *Parus*, are related to protection. I even misdoubt much whether the less blackness of ♀ blackbird is for protection.

Again, can you give me reasons for believing that the moderate differences between the female pheasant, the female *Gallus bankiva*, the female of black grouse, the pea-hen, the female partridge, have all special references to protection under slightly different conditions? I, of course, admit that they are all protected by dull colours, derived, as I think, from some dull-ground progenitor; and I account partly for their difference by partial transference of colour from the male, and by other means too long to specify; but I earnestly wish to see reason to believe that each is specially adapted for concealment to its environment.

I grieve to differ from you, and it actually terrifies me and

* The symbols ♂, ♀, stand for male and female.

makes me constantly distrust myself. I fear we shall never quite understand each other. I value the cases of bright-coloured, incubating male fishes, and brilliant female butterflies, solely as showing that one sex may be made brilliant without any necessary transference of beauty to the other sex; for in these cases I cannot suppose that beauty in the other sex was checked by selection.

I fear this letter will trouble you to read it. A very short answer about your belief in regard to the ♀ finches and gallinaceæ would suffice.

Believe me, my dear Wallace,

Yours very sincerely,

CH. DARWIN.

C. Darwin to J. D. Hooker.

Down, May 25 [1870].

. . . . Last Friday we all went to the Bull Hotel at Cambridge to see the boys, and for a little rest and enjoyment. The backs of the Colleges are simply paradisaical. On Monday I saw Sedgwick, who was most cordial and kind; in the morning I thought his brain was enfeebled; in the evening he was brilliant and quite himself. His affection and kindness charmed us all. My visit to him was in one way unfortunate; for after a long sit he proposed to take me to the museum, and I could not refuse, and in consequence he utterly prostrated me; so that we left Cambridge next morning, and I have not recovered the exhaustion yet. Is it not humiliating to be thus killed by a man of eighty-six, who evidently never dreamed that he was killing me? As he said to me, "Oh, I consider you as a mere baby to me!" I saw Newton several times, and several nice friends of F.'s. But Cambridge without dear Henslow was not itself; I tried to get to the two old houses, but it was too far for me. . . .

*C. Darwin to B. J. Sulivan.**

Down, June 30 [1870].

MY DEAR SULIVAN,—It was very good of you to write to me so long a letter, telling me much about yourself and your children, which I was extremely glad to hear. Think what a benighted wretch I am, seeing no one and reading but little in the newspapers, for I did not know (until seeing the paper of your Natural History Society) that you were a K.C.B. Most heartily glad I am that the Government have at last appreciated your most just claim for this high distinction. On the other hand, I am sorry to hear so poor an account of your health; but you were surely very rash to do all that you did and then pass through so exciting a scene as a ball at the Palace. It was enough to have tired a man in robust health. Complete rest will, however, I hope, quite set you up again. As for myself, I have been rather better of late, and if nothing disturbs me I can do some hours' work every day. I shall this autumn publish another book partly on man, which I dare say many will decry as very wicked. I could have travelled to Oxford, but could no more have withstood the excitement of a commemoration † than I could a ball at Buckingham Palace. Many thanks for your kind remarks about my boys. Thank God, all give me complete satisfaction; my fourth stands second at Woolwich, and will be an Engineer Officer at Christmas. My wife desires to be very kindly remembered to Lady Sulivan, in which I very sincerely join, and in congratulation about your daughter's marriage. We are at present solitary, for all our younger children are

* Admiral Sir James Sulivan was a lieutenant on board the *Beagle*.

† This refers to an invitation to receive the honorary degree of D.C.L. He was one of those nominated for the degree by Lord Salisbury on assuming the office of Chancellor of the University of Oxford. The fact that the honour was declined on the score of ill-health was published in the *Oxford University Gazette*, June 17, 1870.

gone a tour in Switzerland. I had never heard a word about the success of the T. del Fuego mission. It is most wonderful, and shames me, as I always prophesied utter failure. It is a grand success. I shall feel proud if your Committee think fit to elect me an honorary member of your society. With all good wishes and affectionate remembrances of ancient days,

Believe me, my dear Sulivan,

Your sincere friend,

CH. DARWIN.

[My father's connection with the South American Mission, which is referred to in the above letter, has given rise to some public comment, and has been to some extent misunderstood. The Archbishop of Canterbury, speaking at the annual meeting of the South American Missionary Society, April 21st, 1885,* said that the Society "drew the attention of Charles Darwin, and made him, in his pursuit of the wonders of the kingdom of nature, realise that there was another kingdom just as wonderful and more lasting." Some discussion on the subject appeared in the *Daily News* of April 23rd, 24th, 29th, 1885, and finally Admiral Sir James Sulivan, on April 24th, wrote to the same journal, giving a clear account of my father's connection with the Society:—

"Your article in the *Daily News* of yesterday induces me to give you a correct statement of the connection between the South American Missionary Society and Mr. Charles Darwin, my old friend and shipmate for five years. I have been closely connected with the Society from the time of Captain Allen Gardiner's death, and Mr. Darwin had often expressed to me his conviction that it was utterly useless to send Missionaries to such a set of savages as the Fuegians, prob-

* I quote a 'Leaflet,' published by the Society.

ably the very lowest of the human race. I had always replied that I did not believe any human beings existed too low to comprehend the simple message of the Gospel of Christ. After many years, I think about 1869,* but I cannot find the letter, he wrote to me that the recent accounts of the Mission proved to him that he had been wrong and I right in our estimates of the native character, and the possibility of doing them good through Missionaries; and he requested me to forward to the Society an enclosed cheque for £5, as a testimony of the interest he took in their good work. On June 6th, 1874, he wrote: 'I am very glad to hear so good an account of the Fuegians, and it is wonderful.' On June 10th, 1879: 'The progress of the Fuegians is wonderful, and had it not occurred would have been to me quite incredible.' On January 3rd, 1880: 'Your extracts [from a journal] 'about the Fuegians are extremely curious, and have interested me much. I have often said that the progress of Japan was the greatest wonder in the world, but I declare that the progress of Fuegia is almost equally wonderful.' On March 20th, 1881: 'The account of the Fuegians interested not only me, but all my family. It is truly wonderful what you have heard from Mr. Bridges about their honesty and their language. I certainly should have predicted that not all the Missionaries in the world could have done what has been done.' On December 1st, 1881, sending me his annual subscription to the Orphanage at the Mission Station, he wrote: 'Judging from the *Missionary Journal*, the Mission in Tierra del Fuego seems going on quite wonderfully well.' "]

* It seems to have been in 1867.

C. Darwin to John Lubbock.

Down, July 17, 1870.

MY DEAR LUBBOCK,—As I hear that the Census will be brought before the House to-morrow, I write to say how much I hope that you will express your opinion on the desirability of queries in relation to consanguineous marriages being inserted. As you are aware, I have made experiments on the subject during several years; *and it is my clear conviction that there is now ample evidence of the existence of a great physiological law, rendering an enquiry with reference to mankind of much importance. In England and many parts of Europe the marriages of cousins are objected to from their supposed injurious consequences; but this belief rests on no direct evidence. It is therefore manifestly desirable that the belief should either be proved false, or should be confirmed*, so that in this latter case the marriages of cousins might be discouraged. If the proper queries are inserted, the returns would show whether married cousins have in their households on the night of the census as many children as have parents who are not related; and should the number prove fewer, we might safely infer either lessened fertility in the parents, or which is more probable, lessened vitality in the offspring.

It is, moreover, much to be wished that the truth of the often repeated assertion that consanguineous marriages lead to deafness, and dumbness, blindness, &c., should be ascertained; and all such assertions could be easily tested by the returns from a single census.

Believe me,

Yours very sincerely,

CHARLES DARWIN.

[When the Census Act was passing through the House of Commons, Sir John Lubbock and Dr. Playfair attempted to carry out this suggestion. The question came to a division, which was lost, but not by many votes.

The subject of cousin marriages was afterwards investigated by my brother.* The results of this laborious piece of work were negative; the author sums up in the sentence:—

"My paper is far from giving anything like a satisfactory solution of the question as to the effects of consanguineous marriages, but it does, I think, show that the assertion that this question has already been set at rest, cannot be substantiated."]

* "Marriages between First Cousins in England, and their Effects." By George Darwin. 'Journal of the Statistical Society,' June 1875.

CHAPTER IV.

PUBLICATION OF THE 'DESCENT OF MAN.'

THE 'EXPRESSION OF THE EMOTIONS.'

1871–1873.

[THE last revise of the 'Descent of Man' was corrected on January 15th, 1871, so that the book occupied him for about three years. He wrote to Sir J. Hooker: "I finished the last proofs of my book a few days ago; the work half-killed me, and I have not the most remote idea whether the book is worth publishing."

He also wrote to Dr. Gray:—

"I have finished my book on the 'Descent of Man,' &c., and its publication is delayed only by the Index: when published, I will send you a copy, but I do not know that you will care about it. Parts, as on the moral sense, will, I dare say, aggravate you, and if I hear from you, I shall probably receive a few stabs from your polished stiletto of a pen."

The book was published on February 24, 1871. 2500 copies were printed at first, and 5000 more before the end of the year. My father notes that he received for this edition £1470. The letters given in the present chapter deal with its reception, and also with the progress of the work on Expression. The letters are given, approximately, in chronological order, an arrangement which necessarily separates letters of kindred subject-matter, but gives perhaps a truer picture of the mingled interests and labours of my father's life.

Nothing can give a better idea (in a small compass) of the growth of Evolutionism, and its position at this time, than a quotation from Mr. Huxley *:—

"The gradual lapse of time has now separated us by more than a decade from the date of the publication of the 'Origin of Species;' and whatever may be thought or said about Mr. Darwin's doctrines, or the manner in which he has propounded them, this much is certain, that in a dozen years the 'Origin of Species' has worked as complete a revolution in Biological Science as the 'Principia' did in Astronomy;" and it has done so, "because, in the words of Helmholtz, it contains 'an essentially new creative thought.' And, as time has slipped by, a happy change has come over Mr. Darwin's critics. The mixture of ignorance and insolence which at first characterised a large proportion of the attacks with which he was assailed, is no longer the sad distinction of anti-Darwinian criticism."

A passage in the Introduction to the 'Descent of Man' shows that the author recognised clearly this improvement in the position of Evolutionism. "When a naturalist like Carl Vogt ventures to say in his address, as President of the National Institution of Geneva (1869), 'personne, en Europe au moins, n'ose plus soutenir la création indépendante et de toutes pièces, des espèces,' it is manifest that at least a large number of naturalists must admit that species are the modified descendants of other species; and this especially holds good with the younger and rising naturalists. . . . Of the older and honoured chiefs in natural science, many, unfortunately, are still opposed to Evolution in every form."

In Mr. James Hague's pleasantly written article, "A Reminiscence of Mr. Darwin" ('Harper's Magazine,' October 1884),

* 'Contemporary Review,' 1871.

he describes a visit to my father "early in 1871," * shortly after the publication of the 'Descent of Man.' Mr. Hague represents my father as "much impressed by the general assent with which his views had been received," and as remarking that "everybody is talking about it without being shocked."

Later in the year the reception of the book is described in different language in the 'Edinburgh Review': † "On every side it is raising a storm of mingled wrath, wonder and admiration."

With regard to the subsequent reception of the 'Descent of Man,' my father wrote to Dr. Dohrn, February 3, 1872:—

"I did not know until reading your article,‡ that my 'Descent of Man' had excited so much *furore* in Germany. It has had an immense circulation in this country and in America, but has met the approval of hardly any naturalists as far as I know. Therefore I suppose it was a mistake on my part to publish it; but, anyhow, it will pave the way for some better work."

The book on the 'Expression of the Emotions' was begun on January 17th, 1871, the last proof of the 'Descent of Man' having been finished on January 15th. The rough copy was finished by April 27th, and shortly after this (in June) the work was interrupted by the preparation of a sixth edition of the 'Origin.' In November and December the proofs of the 'Expression' book were taken in hand, and occupied him until the following year, when the book was published.

Some references to the work on Expression have occurred in letters already given, showing that the foundation of the book was, to some extent, laid down for some years before he

* It must have been at the end of February, within a week after the publication of the book.

† July 1871. An adverse criticism. The reviewer sums up by saying that: "Never perhaps in the history of philosophy have such wide generalisations been derived from such a small basis of fact."

‡ In 'Das Ausland.'

began to write it. Thus he wrote to Dr. Asa Gray, April 15, 1867:—

"I have been lately getting up and looking over my old notes on Expression, and fear that I shall not make so much of my hobby-horse as I thought I could; nevertheless, it seems to me a curious subject which has been strangely neglected."

It should, however, be remembered that the subject had been before his mind, more or less, from 1837 or 1838, as I judge from entries in his early note-books. It was in December 1839, that he began to make observations on children.

The work required much correspondence, not only with missionaries and others living among savages, to whom he sent his printed queries, but among physiologists and physicians. He obtained much information from Professor Donders, Sir W. Bowman, Sir James Paget, Dr. W. Ogle, Dr. Crichton Browne, as well as from other observers.

The first letter refers to the 'Descent of Man.']

C. Darwin to A. R. Wallace.

Down, January 30 [1871].

MY DEAR WALLACE,—Your note * has given me very great pleasure, chiefly because I was so anxious not to treat you

* In the note referred to, dated January 27, Mr. Wallace wrote:—"Many thanks for your first volume which I have just finished reading through with the greatest pleasure and interest; and I have also to thank you for the great tenderness with which you have treated me and my heresies."

The heresy is the limitation of natural selection as applied to man. My father wrote ('Descent of Man,' i. p. 137):—"I cannot therefore understand how it is that Mr. Wallace maintains that 'natural selection could only have endowed the savage with a brain a little superior to that of an ape.'" In the above quoted letter Mr. Wallace wrote:—"Your chapters on 'Man' are of intense interest, but as touching my special heresy not as yet altogether convincing, though of course I fully agree with every word and every argument which goes to prove the evolution or development of man out of a lower form."

with the least disrespect, and it is so difficult to speak fairly when differing from any one. If I had offended you, it would have grieved me more than you will readily believe. Secondly, I am greatly pleased to hear that Vol. I. interests you; I have got so sick of the whole subject that I felt in utter doubt about the value of any part. I intended, when speaking of females not having been specially modified for protection, to include the prevention of characters acquired by the ♂ being transmitted to ♀; but I now see it would have been better to have said "specially acted on," or some such term. Possibly my intention may be clearer in Vol. II. Let me say that my conclusions are chiefly founded on the consideration of all animals taken in a body, bearing in mind how common the rules of sexual differences appear to be in all classes. The first copy of the chapter on Lepidoptera agreed pretty closely with you. I then worked on, came back to Lepidoptera, and thought myself compelled to alter it—finished Sexual Selection and for the last time went over Lepidoptera, and again I felt forced to alter it. I hope to God there will be nothing disagreeable to you in Vol. II., and that I have spoken fairly of your views; I am fearful on this head, because I have just read (but not with sufficient care) Mivart's book,* and I feel *absolutely certain* that he meant to be fair (but he was stimulated by theological fervour); yet I do not think he has been quite fair. . . . The part which, I think, will have most influence is where he gives the whole series of cases like that of the whalebone, in which we cannot explain the gradational steps; but such cases have no weight on my mind—if a few fish were extinct, who on earth would have ventured even to conjecture that lungs had originated in a swim-bladder? In such a case as the Thylacine, I think he was bound to say that the resemblance of the jaw to that of the dog is superficial; the number and correspondence and development of teeth being widely different. I think again when speaking

* 'The Genesis of Species,' by St. G. Mivart, 1871.

of the necessity of altering a number of characters together, he ought to have thought of man having power by selection to modify simultaneously or almost simultaneously many points, as in making a greyhound or racehorse—as enlarged upon in my 'Domestic Animals.' Mivart is savage or contemptuous about my "moral sense," and so probably will you be. I am extremely pleased that he agrees with my position, *as far as animal nature is concerned*, of man in the series; or if anything, thinks I have erred in making him too distinct.

Forgive me for scribbling at such length. You have put me quite in good spirits; I did so dread having been unintentionally unfair towards your views. I hope earnestly the second volume will escape as well. I care now very little what others say. As for our not quite agreeing, really in such complex subjects, it is almost impossible for two men who arrive independently at their conclusions to agree fully, it would be unnatural for them to do so.

Yours ever, very sincerely,

CH. DARWIN.

[Professor Haeckel seems to have been one of the first to write to my father about the 'Descent of Man.' I quote from his reply:—

"I must send you a few words to thank you for your interesting, and I may truly say, charming letter. I am delighted that you approve of my book, as far as you have read it. I felt very great difficulty and doubt how often I ought to allude to what you have published; strictly speaking every idea, although occurring independently to me, if published by you previously ought to have appeared as if taken from your works, but this would have made my book very dull reading; and I hoped that a full acknowledgment at the beginning would suffice.* I cannot tell you how glad I am to

* In the introduction to the 'Descent of Man' the author wrote:— "This last naturalist [Haeckel] . . . has recently . . . published his 'Na-

find that I have expressed my high admiration of your labours with sufficient clearness; I am sure that I have not expressed it too strongly."]

C. Darwin to A. R. Wallace.

Down, March 16, 1871.

MY DEAR WALLACE,—I have just read your grand review.* It is in every way as kindly expressed towards myself as it is excellent in matter. The Lyells have been here, and Sir C. remarked that no one wrote such good scientific reviews as you, and as Miss Buckley added, you delight in picking out all that is good, though very far from blind to the bad. In all this I most entirely agree. I shall always consider your review as a great honour; and however much my book may hereafter be abused, as no doubt it will be, your review will console me, notwithstanding that we differ so greatly. I will keep your objections to my views in my mind, but I fear that the latter are almost stereotyped in my mind. I thought for long weeks about the inheritance and selection difficulty, and covered quires of paper with notes in trying to get out of it, but could not, though clearly seeing that it would be a great relief if I could. I will confine myself to two or three remarks. I have been much impressed with what you urge against colour,† in the case of insects, having been acquired

türliche Schöpfungs-geschichte,' in which he fully discusses the genealogy of man. If this work had appeared before my essay had been written, I should probably never have completed it. Almost all the conclusions at which I have arrived, I find confirmed by this naturalist, whose knowledge on many points is much fuller than mine."

* *Academy*, March 15, 1871.

† Mr. Wallace says that the pairing of butterflies is probably determined by the fact that one male is stronger-winged, or more pertinacious than the rest, rather than by the choice of the females. He quotes the case of caterpillars which are brightly coloured and yet sexless. Mr. Wallace also makes the good criticism, that the 'Descent of Man' consists of two books mixed together.

through sexual selection. I always saw that the evidence was very weak; but I still think, if it be admitted that the musical instruments of insects have been gained through sexual selection, that there is not the least improbability in colour having been thus gained. Your argument with respect to the denudation of mankind and also to insects, that taste on the part of one sex would have to remain nearly the same during many generations, in order that sexual selection should produce any effect, I agree to; and I think this argument would be sound if used by one who denied that, for instance, the plumes of birds of Paradise had been so gained. I believe you admit this, and if so I do not see how your argument applies in other cases. I have recognised for some short time that I have made a great omission in not having discussed, as far as I could, the acquisition of taste, its inherited nature, and its permanence within pretty close limits for long periods.

[With regard to the success of the 'Descent of Man,' I quote from a letter to Professor Ray Lankester (March 22, 1871):—

"I think you will be glad to hear, as a proof of the increasing liberality of England, that my book has sold wonderfully . . . . and as yet no abuse (though some, no doubt, will come, strong enough), and only contempt even in the poor old *Athenæum*."

As to reviews that struck him he wrote to Mr. Wallace (March 24, 1871):—

"There is a very striking second article on my book in the *Pall Mall*. The articles in the *Spectator** have also interested me much."

* *Spectator*, March 11 and 18, 1871. With regard to the evolution of conscience the reviewer thinks that my father comes much nearer to the "kernel of the psychological problem" than many of his predecessors. The second article contains a good discussion of the bearing of the book on the question of design, and concludes by finding in it a vindication of Theism more wonderful than that in Paley's 'Natural Theology.'

On March 20 he wrote to Mr. Murray:—

"Many thanks for the *Nonconformist* [March 8, 1871]. I like to see all that is written, and it is of some real use. If you hear of reviewers in out-of-the-way papers, especially the religious, as *Record*, *Guardian*, *Tablet*, kindly inform me. It is wonderful that there has been no abuse * as yet, but I suppose I shall not escape. On the whole, the reviews have been highly favourable."

The following extract from a letter to Mr. Murray (April 13, 1871) refers to a review in the *Times*.†

"I have no idea who wrote the *Times* review. He has no knowledge of science, and seems to me a wind-bag full of metaphysics and classics, so that I do not much regard his adverse judgment, though I suppose it will injure the sale."

A review of the 'Descent of Man,' which my father spoke of as "capital," appeared in the *Saturday Review* (Mar. 4 and 11, 1871). A passage from the first notice (Mar. 4) may be quoted in illustration of the broad basis, as regards general acceptance, on which the doctrine of Evolution now stood: "He claims to have brought man himself, his origin and constitution, within that unity which he had previously sought to trace through all lower animal forms. The growth of opinion in the interval, due in chief measure to his own intermediate works, has placed the discussion of this problem

* "I feel a full conviction that my chapter on man will excite attention and plenty of abuse, and I suppose abuse is as good as praise for selling a book."—(From a letter to Mr. Murray, Jan. 31, 1867.)

† *Times*, April 7 and 8, 1871. The review is not only unfavourable as regards the book under discussion, but also as regards Evolution in general, as the following citation will show: "Even had it been rendered highly probable, which we doubt, that the animal creation has been developed into its numerous and widely different varieties by mere evolution, it would still require an independent investigation of overwhelming force and completeness to justify the presumption that man is but a term in this self-evolving series."

in a position very much in advance of that held by it fifteen years ago. The problem of Evolution is hardly any longer to be treated as one of first principles; nor has Mr. Darwin to do battle for a first hearing of his central hypothesis, upborne as it is by a phalanx of names full of distinction and promise, in either hemisphere."

The infolded point of the human ear, discovered by Mr. Woolner, and described in the 'Descent of Man,' seems especially to have struck the popular imagination; my father wrote to Mr. Woolner:—

"The tips to the ears have become quite celebrated. One reviewer ('Nature') says they ought to be called, as I suggested in joke, *Angulus Woolnerianus*.* A German is very proud to find that he has the tips well developed, and I believe will send me a photograph of his ears."]

C. Darwin to John Brodie Innes.†

Down, May 29 [1871].

My dear Innes,—I have been very glad to receive your pleasant letter, for, to tell you the truth, I have sometimes wondered whether you would not think me an outcast and a reprobate after the publication of my last book ['Descent'].‡ I do not wonder at all at your not agreeing with me, for a good many professed naturalists do not. Yet when I see in how extraordinary a manner the judgment of naturalists has changed since I published the 'Origin,' I feel convinced that there will be in ten years quite as much unanimity about man, as far as his corporeal frame is concerned. . . .

* 'Nature,' April 6, 1871. The term suggested is *Angulus Woolnerii.*

† Rev. J. Brodie Innes, of Milton Brodie, formerly Vicar of Down.

‡ In a letter of my father's to Mr. Innes, he says:—"We often differed, but you are one of those rare mortals from whom one can differ and yet feel no shade of animosity, and that is a thing which I should feel very proud of, if any one could say it of me."

[The following letters, addressed to Dr. Ogle, deal with the progress of the work on Expression.]

Down, March 12 [1871].

MY DEAR DR. OGLE,—I have received both your letters, and they tell me all that I wanted to know in the clearest possible way, as, indeed, all your letters have ever done. I thank you cordially. I will give the case of the murderer * in my hobby-horse essay on Expression. I fear that the Eustachian tube question must have cost you a deal of labour; it is quite a complete little essay. It is pretty clear that the mouth is not opened under surprise merely to improve the hearing. Yet why do deaf men generally keep their mouths open? The other day a man here was mimicking a deaf friend, leaning his head forward and sideways to the speaker, with his mouth well open; it was a lifelike representation of a deaf man. Shakespeare somewhere says: 'Hold your breath, listen" or "hark," I forget which. Surprise hurries the breath, and it seems to me one can breathe, at least hurriedly, much quieter through the open mouth than through the nose. I saw the other day you doubted this. As objection is your province at present, I think breathing through the nose ought to come within it likewise, so do pray consider this point, and let me hear your judgment. Consider the nose to be a flower to be fertilised, and then you will make out all about it.† I have had to allude to your paper on 'Sense of Smell;' ‡ is the paging right, namely, 1, 2, 3? If not, I protest by all the gods against the plan followed by some, of having presentation copies falsely paged; and so does Rolleston, as he wrote to me the other day. In haste.

Yours very sincerely,

C. DARWIN.

* 'Expression of the Emotions,' p. 294. The arrest of a murderer in a hospital, as witnessed by Dr. Ogle.

† Dr. Ogle had corresponded with my father on the subject of the fertilisation of flowers.

‡ Medico-chirurg. Trans. liii.

C. Darwin to W. Ogle.

Down, March 25 [1871].

MY DEAR DR. OGLE,—You will think me a horrid bore, but I beg you, *in relation to a new point for observation*, to imagine as well as you can that you suddenly come across some dreadful object, and act with a sudden little start, a *shudder of horror;* please do this once or twice, and observe yourself as well as you can, and *afterwards* read the rest of this note, which I have consequently pinned down. I find, to my surprise, whenever I act thus my platysma contracts. Does yours? (N.B.—See what a man will do for science; I began this note with a horrid fib, namely, that I want you to attend to a new point.*) I will try and get some persons thus to act who are so lucky as not to know that they even possess this muscle, so troublesome for any one making out about expression. Is a shudder akin to the rigor or shivering before fever? If so, perhaps the platysma could be observed in such cases. Paget told me that he had attended much to shivering, and had written in MS. on the subject, and been much perplexed about it. He mentioned that passing a catheter often causes shivering. Perhaps I will write to him about the platysma. He is always most kind in aiding me in all ways, but he is so overworked that it hurts my conscience to trouble him, for I have a conscience, little as you have reason to think so. Help me if you can, and forgive me. Your murderer case has come in splendidly as the acme of prostration from fear.

Yours very sincerely,

CH. DARWIN.

* The point was doubtless described as a new one, to avoid the possibility of Dr. Ogle's attention being directed to the platysma, a muscle which had been the subject of discussion in other letters.

C. Darwin to Dr. Ogle.

Down, April 29 [1871].

MY DEAR DR. OGLE,—I am truly obliged for all the great trouble which you have so kindly taken. I am sure you have no cause to say that you are sorry you can give me no definite information, for you have given me far more than I ever expected to get. The action of the platysma is not very important for me, but I believe that you will fully understand (for I have always fancied that our minds were very similar) the intolerable desire I had not to be utterly baffled. Now I know that it sometimes contracts from fear and from shuddering, but not apparently from a prolonged state of fear such as the insane suffer. . . .

[Mr. Mivart's 'Genesis of Species,'—a contribution to the literature of Evolution, which excited much attention,—was published in 1871, before the appearance of the 'Descent of Man.' To this book the following letter (June 21, 1871) from the late Chauncey Wright * to my father, refers:—

"I send . . . revised proofs of an article which will be published in the July number of the 'North American Review,' sending it in the hope that it will interest or even be of greater value to you. Mr. Mivart's book ['Genesis of Species'] of which this article is substantially a review, seems to me a very good background from which to present the considerations which I have endeavoured to set forth in the article, in defence and illustration of the theory of Natural

* Chauncey Wright was born at Northampton, Massachusetts, Sept. 20, 1830, and came of a family settled in that town since 1654. He became in 1852 a computer in the Nautical Almanac office at Cambridge, Mass., and lived a quiet uneventful life, supported by the small stipend of his office, and by what he earned from his occasional articles, as well by a little teaching. He thought and read much on metaphysical subjects, but on the whole with an outcome (as far as the world was concerned) not commensurate to the power of his mind. He seems to have been a man of strong individuality, and to have made a lasting impression on his friends. He died in Sept. 1875.

Selection. My special purpose has been to contribute to the theory by placing it in its proper relations to philosophical inquiries in general." *

With regard to the proofs received from Mr. Wright, my father wrote to Mr. Wallace :]

Down, July 9 [1871].

MY DEAR WALLACE,—I send by this post a review by Chauncey Wright, as I much want your opinion of it as soon as you can send it. I consider you an incomparably better critic than I am. The article, though not very clearly written, and poor in parts from want of knowledge, seems to me admirable. Mivart's book is producing a great effect against Natural Selection, and more especially against me. Therefore if you think the article even somewhat good I will write and get permission to publish it as a shilling pamphlet, together with the MS. additions (enclosed), for which there was not room at the end of the review. . . .

I am now at work at a new and cheap edition of the 'Origin,' and shall answer several points in Mivart's book, and introduce a new chapter for this purpose ; but I treat the subject so much more concretely, and I dare say less philosophically, than Wright, that we shall not interfere with each other. You will think me a bigot when I say, after studying Mivart, I was never before in my life so convinced of the *general* (*i.e.* not in detail) truth of the views in the 'Origin.' I grieve to see the omission of the words by Mivart, detected by Wright. † I complained to Mivart that in two cases he quotes only the commencement of sentences by me, and thus

* 'Letters of Chauncey Wright,' by J. B. Thayer. Privately printed, 1878, p. 230.

† 'North American Review' vol. 113, pp. 83, 84. Chauncey Wright points out that the words omitted are "essential to the point on which he [Mr. Mivart] cites Mr. Darwin's authority." It should be mentioned that the passage from which words are omitted is not given within inverted commas by Mr. Mivart.

modifies my meaning; but I never supposed he would have omitted words. There are other cases of what I consider unfair treatment. I conclude with sorrow that though he means to be honourable, he is so bigoted that he cannot act fairly. . . .

C. Darwin to Chauncey Wright.

Down, July 14, 1871.

MY DEAR SIR,—I have hardly ever in my life read an article which has given me so much satisfaction as the review which you have been so kind as to send me. I agree to almost everything which you say. Your memory must be wonderfully accurate, for you know my works as well as I do myself, and your power of grasping other men's thoughts is something quite surprising; and this, as far as my experience goes, is a very rare quality. As I read on I perceived how you have acquired this power, viz. by thoroughly analyzing each word.

. . . Now I am going to beg a favour. Will you provisionally give me permission to reprint your article as a shilling pamphlet? I ask only provisionally, as I have not yet had time to reflect on the subject. It would cost me, I fancy, with advertisements, some £20 or £30; but the worst is that, as I hear, pamphlets never will sell. And this makes me doubtful. Should you think it too much trouble to send me a title *for the chance?* The title ought, I think, to have Mr. Mivart's name on it.

. . . If you grant permission and send a title, you will kindly understand that I will first make further enquiries whether there is any chance of a pamphlet being read.

Pray believe me yours very sincerely obliged,

CH. DARWIN.

[The pamphlet was published in the autumn, and on October 23 my father wrote to Mr. Wright:—

"It pleases me much that you are satisfied with the appearance of your pamphlet. I am sure it will do our cause good service; and this same opinion Huxley has expressed to me. ('Letters of Chauncey Wright,' p. 235.)"]

C. Darwin to A. R. Wallace.

Down, July 12 [1871].

. . . . I feel very doubtful how far I shall succeed in answering Mivart, it is so difficult to answer objections to doubtful points, and make the discussion readable. I shall make only a selection. The worst of it is, that I cannot possibly hunt through all my references for isolated points, it would take me three weeks of intolerably hard work. I wish I had your power of arguing clearly. At present I feel sick of everything, and if I could occupy my time and forget my daily discomforts, or rather miseries, I would never publish another word. But I shall cheer up, I dare say, soon, having only just got over a bad attack. Farewell; God knows why I bother you about myself. I can say nothing more about missing-links than what I have said. I should rely much on pre-silurian times; but then comes Sir W. Thomson like an odious spectre. Farewell.

. . . There is a most cutting review of me in the 'Quarterly';* I have only read a few pages. The skill and style make me think of Mivart. I shall soon be viewed as the most despicable of men. This 'Quarterly Review' tempts me to republish Ch. Wright, even if not read by any one, just to show some one will say a word against Mivart, and that his (*i.e.* Mivart's) remarks ought not to be swallowed without some reflection. . . . God knows whether my strength and spirit will last out to write a chapter versus Mivart and others; I do so hate controversy and feel I shall do it so badly.

* July 1871.

[The above-mentioned 'Quarterly' review was the subject of an article by Mr. Huxley in the November number of the 'Contemporary Review.' Here, also, are discussed Mr. Wallace's 'Contribution to the Theory of Natural Selection,' and the second edition of Mr. Mivart's 'Genesis of Species.' What follows is taken from Mr. Huxley's article. The 'Quarterly' reviewer, though being to some extent an evolutionist, believes that Man "differs more from an elephant or a gorilla, than do these from the dust of the earth on which they tread." The reviewer also declares that my father has "with needless opposition, set at naught the first principles of both philosophy and religion." Mr. Huxley passes from the 'Quarterly' reviewer's further statement, that there is no necessary opposition between evolution and religion, to the more definite position taken by Mr. Mivart, that the orthodox authorities of the Roman Catholic Church agree in distinctly asserting derivative creation, so that "their teachings harmonize with all that modern science can possibly require." Here Mr. Huxley felt the want of that "study of Christian philosophy" (at any rate, in its Jesuitic garb), which Mr. Mivart speaks of, and it was a want he at once set to work to fill up. He was then staying at St. Andrews, whence he wrote to my father:—

"By great good luck there is an excellent library here, with a good copy of Suarez,* in a dozen big folios. Among these I dived, to the great astonishment of the librarian, and looking into them 'as the careful robin eyes the delver's toil' (*vide* 'Idylls'), I carried off the two venerable clasped volumes which were most promising." Even those who know Mr. Huxley's unrivalled power of tearing the heart out of a book must marvel at the skill with which he has made Suarez speak on his side. "So I have come out," he wrote, "in the new character of a defender of Catholic orthodoxy, and upset Mivart out of the mouth of his own prophet."

* The learned Jesuit on whom Mr. Mivart mainly relies.

The remainder of Mr. Huxley's critique is largely occupied with a dissection of the 'Quarterly' reviewer's psychology, and his ethical views. He deals, too, with Mr. Wallace's objections to the doctrine of Evolution by natural causes when applied to the mental faculties of Man. Finally, he devotes a couple of pages to justifying his description of the 'Quarterly' reviewer's "treatment of Mr. Darwin as alike unjust and unbecoming."

It will be seen that the two following letters were written before the publication of Mr. Huxley's article.]

C. Darwin to T. H. Huxley.

Down, September 21 [1871].

MY DEAR HUXLEY,—Your letter has pleased me in many ways, to a wonderful degree. . . . What a wonderful man you are to grapple with those old metaphysico-divinity books. It quite delights me that you are going to some extent to answer and attack Mivart. His book, as you say, has produced a great effect; yesterday I perceived the reverberations from it, even from Italy. It was this that made me ask Chauncey Wright to publish at my expense his article, which seems to me very clever, though ill-written. He has not knowledge enough to grapple with Mivart in detail. I think there can be no shadow of doubt that he is the author of the article in the 'Quarterly Review' . . . I am preparing a new edition of the 'Origin,' and shall introduce a new chapter in answer to miscellaneous objections, and shall give up the greater part to answer Mivart's cases of difficulty of incipient structures being of no use: and I find it can be done easily. He never states his case fairly, and makes wonderful blunders. . . . The pendulum is now swinging against our side, but I feel positive it will soon swing the other way; and no mortal man will do half as much as you in giving it a start in the right direction, as you did at the first commencement. God forgive me for writing so long and egotistical a letter; but it is your fault, for you have so delighted me ; I never dreamed that you would have time to say a word in defence of the cause which you have so often defended. It will be a long battle, after we are dead and gone. . . . Great is the power of misrepresentation. . . .

C. Darwin to T. H. Huxley.

Down, September 30 [1871].

MY DEAR HUXLEY,—It was very good of you to send the proof-sheets, for I was *very* anxious to read your article. I have been delighted with it. How you do smash Mivart's theology : it is almost equal to your article versus Comte,—* that never can be transcended. . . . But I have been pre-eminently glad to read your discussion on [the 'Quarterly' reviewer's] metaphysics, especially about reason and his definition of it. I felt sure he was wrong, but having only common observation and sense to trust to, I did not know what to say in my second edition of my 'Descent.' Now a footnote and reference to you will do the work. . . . For me, this is one of the most *important* parts of the review. But for *pleasure*, I have been particularly glad that my few words † on the distinction, if it can be so called, between Mivart's two forms of morality, caught your attention. I am so pleased that you take the same view, and give authorities for it ; but I searched Mill in vain on this head. How well you argue the whole case. I am mounting climax on climax ; for after all there is nothing, I think, better in your whole review than your

* 'Fortnightly Review,' 1869. With regard to the relations of Positivism to Science, my father wrote to Mr. Spencer in 1875 : "How curious and amusing it is to see to what an extent the Positivists hate all men of science ; I fancy they are dimly conscious what laughable and gigantic blunders their prophet made in predicting the course of science."

† 'Descent of Man,' vol. i. p. 87. A discussion on the question whether an act done impulsively or instinctively can be called moral.

arguments *v.* Wallace on the intellect of savages. I must tell you what Hooker said to me a few years ago. "When I read Huxley, I feel quite infantile in intellect." By Jove I have felt the truth of this throughout your review. What a man you are. There are scores of splendid passages, and vivid flashes of wit. I have been a good deal more than merely pleased by the concluding part of your review; and all the more, as I own I felt mortified by the accusation of bigotry, arrogance, &c., in the 'Quarterly Review.' But I assure you, he may write his worst, and he will never mortify me again.

My dear Huxley, yours gratefully,

CHARLES DARWIN.

C. Darwin to F. Müller.

Haredene, Albury, August 2 [1871].

MY DEAR SIR,—Your last letter has interested me greatly; it is wonderfully rich in facts and original thoughts. First, let me say that I have been much pleased by what you say about my book. It has had a *very large* sale; but I have been much abused for it, especially for the chapter on the moral sense; and most of my reviewers consider the book as a poor affair. God knows what its merits may really be; all that I know is that I did my best. With familiarity I think naturalists will accept sexual selection to a greater extent than they now seem inclined to do. I should very much like to publish your letter, but I do not see how it could be made intelligible, without numerous coloured illustrations, but I will consult Mr. Wallace on this head. I earnestly hope that you keep notes of all your letters and that some day you will publish a book: 'Notes of a Naturalist in S. Brazil,' or some such title. Wallace will hardly admit the possibility of sexual selection with Lepidoptera, and no doubt it is very improbable. Therefore, I am very glad to hear of your cases (which I will quote in the next edition) of the two sets of

Hesperiadæ, which display their wings differently, according to which surface is coloured. I cannot believe that such display is accidental and purposeless. . . .

No fact of your letter has interested me more than that about mimicry. It is a capital fact about the males pursuing the wrong females. You put the difficulty of the first steps in imitation in a most striking and *convincing* manner. Your idea of sexual selection having aided protective imitation interests me greatly, for the same idea had occurred to me in quite different cases, viz. the dulness of all animals in the Galapagos Islands, Patagonia, &c., and in some other cases; but I was afraid even to hint at such an idea. Would you object to my giving some such sentence as follows: "F Müller suspects that sexual selection may have come into play, in aid of protective imitation, in a very peculiar manner, which will appear extremely improbable to those who do not fully believe in sexual selection. It is that the appreciation of certain colour is developed in those species which frequently behold other species thus ornamented." Again let me thank you cordially for your most interesting letter. . . .

*C. Darwin to E. B. Tylor.**

Down [Sept. 24, 1871].

MY DEAR SIR,—I hope that you will allow me to have the pleasure of telling you how greatly I have been interested by your 'Primitive Culture,' now that I have finished it. It seems to me a most profound work, which will be certain to have permanent value, and to be referred to for years to come. It is wonderful how you trace animism from the lower races up to the religious belief of the highest races. It will make me for the future look at religion—a belief in the soul, &c.—from a new point of view. How curious, also, are the survivals or

* Keeper of the Museum, and Reader in Anthropology at Oxford.

rudiments of old customs. . . . You will perhaps be surprised at my writing at so late a period, but I have had the book read aloud to me, and from much ill-health of late, could only stand occasional short reads. The undertaking must have cost you gigantic labour. Nevertheless, I earnestly hope that you may be induced to treat morals in the same enlarged yet careful manner, as you have animism. I fancy from the last chapter that you have thought of this. No man could do the work so well as you, and the subject assuredly is a most important and interesting one. You must now possess references which would guide you to a sound estimation of the morals of savages; and how writers like Wallace, Lubbock, &c. &c., do differ on this head. Forgive me for troubling you, and believe me, with much respect,

Yours very sincerely,

CH. DARWIN.

1872.

[At the beginning of the year the sixth edition of the 'Origin,' which had been begun in June 1871, was nearly completed. The last sheet was revised on January 10, 1872, and the book was published in the course of the month. This volume differs from the previous ones in appearance and size—it consists of 458 pp. instead of 596 pp., and is a few ounces lighter; it is printed on bad paper, in small type, and with the lines unpleasantly close together. It had, however, one advantage over the previous editions, namely that it was issued at a lower price. It is to be regretted that this the final edition of the 'Origin' should have appeared in so unattractive a form; a form which has doubtless kept many readers from the book.

The discussion suggested by the 'Genesis of Species' was perhaps the most important addition to the book. The objection that incipient structures cannot be of use, was dealt with in some detail, because it seemed to the author that this was the point in Mr. Mivart's book which had struck most readers in England.

It is a striking proof of how wide and general had become the acceptance of his views, that my father found it necessary to insert (sixth edition, p. 424), the sentence: "As a record of a former state of things, I have retained in the foregoing paragraphs and also elsewhere, several sentences which imply that naturalists believe in the separate creation of each species; and I have been much censured for having thus expressed myself. But undoubtedly this was the general belief when the first edition of the present work appeared. . . Now things are wholly changed, and almost every naturalist admits the great principle of evolution."

A small correction introduced into this sixth edition is connected with one of his minor papers: "Note on the habits of the Pampas Woodpecker." * The paper in question was a reply to Mr. Hudson's remarks on the woodpecker in a previous number of the same journal. The last sentence of my father's paper is worth quoting for its temperate tone: "Finally, I trust that Mr. Hudson is mistaken when he says that any one acquainted with the habits of this bird might be induced to believe that I 'had purposely wrested the truth' in order to prove my theory. He exonerates me from this charge; but I should be loath to think that there are many naturalists who, without any evidence, would accuse a fellow-worker of telling a deliberate falsehood to prove his theory." In the fifth edition of the 'Origin,' p. 220, he wrote:—

"Yet as I can assert not only from my own observation, but from that of the accurate Azara, it [the ground woodpecker] never climbs a tree." In the sixth edition, p. 142, the passage runs "in certain large districts it does not climb trees." And he goes on to give Mr. Hudson's statement, that in other regions it does frequent trees.

* Zoolog. Soc. Proc. 1870.

One of the additions in the sixth edition (p. 149), was a reference to Mr. A. Hyatt's and Professor Cope's theory of "acceleration." With regard to this he wrote (October 10, 1872) in characteristic words to Mr. Hyatt:—

"Permit me to take this opportunity to express my sincere regret at having committed two grave errors in the last edition of my 'Origin of Species,' in my allusion to yours and Professor Cope's views on acceleration and retardation of development. I had thought that Professor Cope had preceded you; but I now well remember having formerly read with lively interest, and marked, a paper by you somewhere in my library, on fossil Cephalopods with remarks on the subject. It seems also that I have quite misrepresented your joint view. This has vexed me much. I confess that I have never been able to grasp fully what you wish to show, and I presume that this must be owing to some dulness on my part."

The sixth edition of the 'Origin' being intended as a popular one, was made to include a glossary of technical terms, "given because several readers have complained . . . that some of the terms used were unintelligible to them." The glossary was compiled by Mr. Dallas, and being an excellent collection of clear and sufficient definitions, must have proved useful to many readers.]

C. Darwin to J. L. A. de Quatrefages.

Down, January 15, 1872.

MY DEAR SIR,—I am much obliged for your very kind letter and exertions in my favour. I had thought that the publication of my last book ['Descent of Man'] would have destroyed all your sympathy with me, but though I estimated very highly your great liberality of mind, it seems that I underrated it.

I am gratified to hear that M. Lacaze-Duthiers will vote for me,* for I have long honoured his name. I cannot help regretting that you should expend your valuable time in trying to obtain for me the honour of election, for I fear, judging from the last time, that all your labour will be in vain. Whatever the result may be, I shall always retain the most lively recollection of your sympathy and kindness, and this will quite console me for my rejection.

With much respect and esteem, I remain, dear Sir,

Yours truly obliged,

CHARLES DARWIN.

P.S.—With respect to the great stress which you lay on man walking on two legs, whilst the quadrumana go on all fours, permit me to remind you that no one much values the great difference in the mode of locomotion, and consequently in structure, between seals and the terrestrial carnivora, or between the almost biped kangaroos and other marsupials.

C. Darwin to August Weismann.†

Down, April 5, 1872.

MY DEAR SIR,—I have now read your essay ‡ with very great interest. Your view of the origin of local races through "Amixie," is altogether new to me, and seems to throw an important light on an obscure problem. There is, however, something strange about the periods or endurance of variability. I formerly endeavoured to investigate the subject, not by looking to past time, but to species of the same genus widely distributed; and I found in many cases that all the species, with perhaps one or two exceptions, were variable. It would be a very interesting subject for a con-

* He was not elected as a corresponding member of the French Academy until 1878.

† Professor of Zoology in Freiburg.

‡ 'Ueber den Einfluss der Isolirung auf die Artbildung.' Leipzig, 1872.

chologist to investigate, viz.: whether the species of the same genus were variable during many successive geological formations. I began to make enquiries on this head, but failed in this, as in so many other things, from the want of time and strength. In your remarks on crossing, you do not, as it seems to me, lay nearly stress enough on the increased vigour of the offspring derived from parents which have been exposed to different conditions. I have during the last five years been making experiments on this subject with plants, and have been astonished at the results, which have not yet been published.

In the first part of your essay, I thought that you wasted (to use an English expression) too much powder and shot on M. Wagner;* but I changed my opinion when I saw how admirably you treated the whole case, and how well you used the facts about the Planorbis. I wish I had studied this latter case more carefully. The manner in which, as you show, the different varieties blend together and make a constant whole, agrees perfectly with my hypothetical illustrations.

Many years ago the late E. Forbes described three closely consecutive beds in a secondary formation, each with representative forms of the same fresh-water shells: the case is evidently analogous with that of Hilgendorf,† but the interesting connecting varieties or links were here absent. I rejoice to think that I formerly said as emphatically as I could, that neither isolation nor time by themselves do anything for the modification of species. Hardly anything in your essay has pleased me so much personally, as to find that you believe to a certain extent in sexual selection. As far as I can judge,

* Prof. Wagner has written two essays on the same subject. 'Die Darwin'sche Theorie und das Migrationsgesetz,' in 1868, and 'Ueber den Einfluss der Geographischen Isolirung, &c.', an address to the Bavarian Academy of Sciences at Munich, 1870.

† "Ueber *Planorbis multiformis* im Steinheimer Süsswasser-kalk." 'Monatsbericht' of the Berlin Academy, 1866.

very few naturalists believe in this. I may have erred on many points, and extended the doctrine too far, but I feel a strong conviction that sexual selection will hereafter be admitted to be a powerful agency. I cannot agree with what you say about the taste for beauty in animals not easily varying. It may be suspected that even the habit of viewing differently coloured surrounding objects would influence their taste, and Fritz Müller even goes so far as to believe that the sight of gaudy butterflies might influence the taste of distinct species. There are many remarks and statements in your essay which have interested me greatly, and I thank you for the pleasure which I have received from reading it.

With sincere respect, I remain,

My dear Sir, yours very faithfully,

CHARLES DARWIN.

P.S.—If you should ever be induced to consider the whole doctrine of sexual selection, I think that you will be led to the conclusion, that characters thus gained by one sex are very commonly transferred in a greater or less degree to the other sex.

[With regard to Moritz Wagner's first Essay, my father wrote to that naturalist, apparently in 1868 :]

DEAR AND RESPECTED SIR,—I thank you sincerely for sending me your 'Migrationsgesetz, &c.,' and for the very kind and most honourable notice which you have taken of my works. That a naturalist who has travelled into so many and such distant regions, and who has studied animals of so many classes, should, to a considerable extent, agree with me, is, I can assure you, the highest gratification of which I am capable. . . . Although I saw the effects of isolation in the case of islands and mountain-ranges, and knew of a few instances of rivers, yet the greater number of your facts were quite unknown to me. I now see that from the want of knowledge I did not make nearly sufficient use of the views which you advocate ; and I almost wish I could believe in its importance to the same extent with you ; for you well show, in a manner which never occurred to me, that it removes many difficulties and objections. But I must still believe that in many large areas all the individuals of the same species have been slowly modified, in the same manner, for instance, as the English race-horse has been improved, that is by the continued selection of the fleetest individuals, without any separation. But I admit that by this process two or more new species could hardly be found within the same limited area ; some degree of separation, if not indispensable, would be highly advantageous ; and here your facts and views will be of great value. . . .

[The following letter bears on the same subject. It refers to Professor M. Wagner's Essay, published in *Das Ausland*, May 31, 1875 :]

C. Darwin to Moritz Wagner.

Down, October 13, 1876.

DEAR SIR,—I have now finished reading your essays, which have interested me in a very high degree, notwithstanding that I differ much from you on various points. For instance, several considerations make me doubt whether species are much more variable at one period than at another, except through the agency of changed conditions. I wish, however, that I could believe in this doctrine, as it removes many difficulties. But my strongest objection to your theory is that it does not explain the manifold adaptations in structure in every organic being—for instance in a Picus for climbing trees and catching insects—or in a Strix for catching animals at night, and so on *ad infinitum*. No theory is in the least satisfactory to me unless it clearly explains such adaptations. I think that you misunderstand my views on isolation. I believe that all the individuals of a species can be slowly modified within the same district, in nearly the same manner as man effects by what I have called the process of unconscious selection. . . . I do not believe that one species will give birth to two or more new species, as long as they are mingled together within the same district. Nevertheless I cannot doubt that many new species have been simultaneously developed within the same large continental area; and in my 'Origin of Species' I endeavoured to explain how two new species might be developed, although they met and intermingled on the *borders* of their range. It would have been a strange fact if I had overlooked the importance of isolation, seeing that it was such cases as that of the Galapagos Archipelago, which chiefly led me to study the origin of species. In my opinion the greatest error which I have committed, has been not allowing sufficient weight to the direct action of the environment, *i.e.* food, climate, &c., independently of natural selection. Modifications thus caused, which are neither of advantage nor disadvantage to the modified organism, would be especially favoured, as I can now see chiefly through your observations, by isolation in a small area, where only a few individuals lived under nearly uniform conditions.

When I wrote the 'Origin,' and for some years afterwards, I could find little good evidence of the direct action of the environment; now there is a large body of evidence, and your case of the Saturnia is one of the most remarkable of which I have heard. Although we differ so greatly, I hope that you will permit me to express my respect for your long-continued and successful labours in the good cause of natural science.

I remain, dear Sir, yours very faithfully,

CHARLES DARWIN.

[The two following letters are also of interest as bearing

on my father's views on the action of isolation as regards the origin of new species:]

C. Darwin to K. Semper.

Down, November 26, 1878.

My dear Professor Semper,—When I published the sixth edition of the 'Origin,' I thought a good deal on the subject to which you refer, and the opinion therein expressed was my deliberate conviction. I went as far as I could, perhaps too far, in agreement with Wagner; since that time I have seen no reason to change my mind, but then I must add that my attention has been absorbed on other subjects. There are two different classes of cases, as it appears to me, viz. those in which a species becomes slowly modified in the same country (of which I cannot doubt there are innumerable instances) and those cases in which a species splits into two or three or more new species; and in the latter case, I should think nearly perfect separation would greatly aid in their "specification," to coin a new word.

I am very glad that you are taking up this subject, for you will be sure to throw much light on it. I remember well, long ago, oscillating much; when I thought of the Fauna and Flora of the Galapagos Islands I was all for isolation, when I thought of S. America I doubted much. Pray believe me,

Yours very sincerely,

Ch. Darwin.

P.S.—I hope that this letter will not be quite illegible, but I have no amanuensis at present.

C. Darwin to K. Semper.

Down, November 30, 1878.

Dear Professor Semper,—Since writing I have recalled some of the thoughts and conclusions which have passed through my mind of late years. In North America, in going from north to south or from east to west, it is clear that the changed conditions of life have modified the organisms in the different regions, so that they now form distinct races or even species. It is further clear that in isolated districts, however small, the inhabitants almost always get slightly modified, and how far this is due to the nature of the slightly different conditions to which they are exposed, and how far to mere interbreeding, in the manner explained by Weismann, I can form no opinion. The same difficulty occurred to me (as shown in my 'Variation of Animals and Plants under Domestication') with respect to the aboriginal breeds of cattle, sheep, &c., in the separated districts of Great Britain, and indeed throughout Europe. As our knowledge advances, very slight differences, considered by systematists as of no importance in structure, are continually found to be functionally important; and I have been especially struck with this fact in the case of plants to which my observations have of late years been confined. Therefore it seems to me rather rash to consider the slight differences between representative species, for instance those inhabiting the different islands of the same archipelago, as of no functional importance, and as not in any way due to natural selection. With respect to all adopted structures, and these are innumerable, I cannot see how M. Wagner's view throws any light, nor indeed do I see at all more clearly than I did before, from the numerous cases which he has brought forward, how and why it is that a long isolated form should almost always become slightly modified. I do not know whether you will care about hearing my further opinion on the point in question, for as before remarked I have not attended much of late years to such questions, thinking it prudent, now that I am growing old, to work at easier subjects.

Believe me, yours very sincerely,

CH. DARWIN.

I hope and trust that you will throw light on these points.

P.S.—I will add another remark which I remember occurred to me when I first read M. Wagner. When a species first arrives on a small island, it will probably increase rapidly, and unless all the individuals change instantaneously (which is improbable in the highest degree), the slowly, more or less, modifying offspring must intercross one with another, and with their unmodified parents, and any offspring not as yet modified. The case will then be like that of domesticated animals which have slowly become modified, either by the action of the external conditions or by the process which I have called the *unconscious selection* by man—*i.e.*, in contrast with methodical selection.

[The letters continue the history of the year 1872, which has been interrupted by a digression on Isolation.]

C. Darwin to the Marquis de Saporta.

Down, April 8, 1872.

DEAR SIR,—I thank you very sincerely and feel much honoured by the trouble which you have taken in giving me your reflections on the origin of Man. It gratifies me extremely that some parts of my work have interested you, and that we agree on the main conclusion of the derivation of man from some lower form.

I will reflect on what you have said, but I cannot at present give up my belief in the close relationship of Man to the higher Simiæ. I do not put much trust in any single character, even that of dentition ; but I put the greatest faith in resemblances in many parts of the whole organisation, for I cannot believe that such resemblances can be due to any cause except close blood relationship. That man is closely allied to the higher Simiæ is shown by the classification of

Linnæus, who was so good a judge of affinity. The man who in England knows most about the structure of the Simiæ, namely, Mr. Mivart, and who is bitterly opposed to my doctrines about the derivation of the mental powers, yet has publicly admitted that I have not put man too close to the higher Simiæ, as far as bodily structure is concerned. I do not think the absence of reversions of structure in man is of much weight; C. Vogt, indeed, argues that [the existence of] Micro-cephalous idiots is a case of reversion. No one who believes in Evolution will doubt that the Phocæ are descended from some terrestrial Carnivore. Yet no one would expect to meet with any such reversion in them. The lesser divergence of character in the races of man in comparison with the species of Simiadæ may perhaps be accounted for by man having spread over the world at a much later period than did the Simiadæ. I am fully prepared to admit the high antiquity of man; but then we have evidence, in the Dryopithecus, of the high antiquity of the Anthropomorphous Simiæ.

I am glad to hear that you are at work on your fossil plants, which of late years have afforded so rich a field for discovery. With my best thanks for your great kindness, and with much respect, I remain,

Dear Sir, yours very faithfully,

CHARLES DARWIN.

[In April, 1872, he was elected to the Royal Society of Holland, and wrote to Professor Donders:—

"Very many thanks for your letter. The honour of being elected a foreign member of your Royal Society has pleased me much. The sympathy of his fellow workers has always appeared to me by far the highest reward to which any scientific man can look. My gratification has been not a little increased by first hearing of the honour from you."]

C. Darwin to Chauncey Wright.

Down, June 3, 1872.

MY DEAR SIR,—Many thanks for your article * in the 'North American Review,' which I have read with great interest. Nothing can be clearer than the way in which you discuss the permanence or fixity of species. It never occurred to me to suppose that any one looked at the case as it seems Mr. Mivart does. Had I read his answer to you, perhaps I should have perceived this; but I have resolved to waste no more time in reading reviews of my works or on Evolution, excepting when I hear that they are good and contain new matter. . . . It is pretty clear that Mr. Mivart has come to the end of his tether on this subject.

As your mind is so clear, and as you consider so carefully the meaning of words, I wish you would take some incidental occasion to consider when a thing may properly be said to be effected by the will of man. I have been led to the wish by reading an article by your Professor Whitney *versus* Schleicher. He argues, because each step of change in language is made by the will of man, the whole language so changes; but I do not think that this is so, as man has no intention or wish to change the language. It is a parallel case with what I have called "unconscious selection," which depends on men consciously preserving the best individuals, and thus unconsciously altering the breed.

My dear Sir, yours sincerely,

CHARLES DARWIN.

[Not long afterwards (September) Mr. Chauncey Wright paid

* The proof-sheets of an article which appeared in the July number of the 'North American Review.' It was a rejoinder to Mr. Mivart's reply ('N. Am. Review,' April 1872) to Mr. Chauncey Wright's pamphlet. Chauncey Wright says of it ('Letters,' p. 238) :—"It is not properly a rejoinder but a new article, repeating and expounding some of the points of my pamphlet, and answering some of Mr. Mivart's replies incidentally."

a visit to Down,* which he described in a letter † to Miss S. Sedgwick (now Mrs. William Darwin): "If you can imagine me enthusiastic—absolutely and unqualifiedly so, without a *but* or criticism, then think of my last evening's and this morning's talks with Mr. Darwin. . . . I was never so worked up in my life, and did not sleep many hours under the hospitable roof. . . . It would be quite impossible to give by way of report any idea of these talks before and at and after dinner, at breakfast, and at leave-taking; and yet I dislike the egotism of 'testifying' like other religious enthusiasts without any verification, or hint of similar experience."]

C. Darwin to Herbert Spencer.

Bassett, Southampton, June 10 [1872].

DEAR SPENCER,—I dare say you will think me a foolish fellow, but I cannot resist the wish to express my unbounded admiration of your article ‡ in answer to Mr. Martineau. It is, indeed, admirable, and hardly less so your second article on Sociology (which, however, I have not yet finished): I never believed in the reigning influence of great men on the world's progress; but if asked why I did not believe, I should have been sorely perplexed to have given a good answer. Every one with eyes to see and ears to hear (the number, I

* Mr. and Mrs. C. L. Brace, who had given much of their lives to philanthropic work in New York, also paid a visit at Down in this summer. Some of their work is recorded in Mr. Brace's 'The Dangerous Classes of New York,' and of this book my father wrote to the author:—

"Since you were here my wife has read aloud to me more than half of your work, and it has interested us both in the highest degree, and we shall read every word of the remainder. The facts seem to me very well told, and the inferences very striking. But after all, this is but a weak part of the impression left on our minds by what we have read; for we are both filled with earnest admiration at the heroic labours of yourself and others."

† 'Letters,' p. 246–248.

‡ "Mr. Martineau on Evolution," by Herbert Spencer, 'Contemporary Review,' July 1872.

fear, are not many) ought to bow their knee to you, and I for one do.

Believe me, yours most sincerely,
C. DARWIN.

C. Darwin to J. D. Hooker.

Down, July 12 [1872].

MY DEAR HOOKER,—I must exhale and express my joy at the way in which the newspapers have taken up your case. I have seen the *Times*, the *Daily News*, and the *Pall Mall*, and hear that others have taken up the case.

The Memorial has done great good this way, whatever may be the result in the action of our wretched Government. On my soul, it is enough to make one turn into an old honest Tory. . . .

If you answer this, I shall be sorry that I have relieved my feelings by writing.

Yours affectionately,
C. DARWIN.

[The memorial here referred to was addressed to Mr. Gladstone, and was signed by a number of distinguished men, including Sir Charles Lyell, Mr. Bentham, Mr. Huxley, and Sir James Paget. It gives a complete account of the arbitrary and unjust treatment received by Sir J. D. Hooker at the hands of his official chief, the First Commissioner of Works. The document is published in full in 'Nature' (July 11, 1872), and is well worth studying as an example of the treatment which it is possible for science to receive from officialism. As 'Nature' observes, it is a paper which must be read with the greatest indignation by scientific men in every part of the world, and with shame by all Englishmen. The signatories of the memorial conclude by protesting against the expected consequences of Sir Joseph Hooker's persecution—namely his resignation, and the loss of "a man honoured for his integrity, beloved for his courtesy and kindliness of heart; and who has spent in the public service not only a stainless but an illustrious life."

Happily this misfortune was averted, and Sir Joseph was freed from further molestation.]

C. Darwin to A. R. Wallace.

Down, August 3 [1872].

MY DEAR WALLACE,—I hate controversy, chiefly perhaps because I do it badly; but as Dr. Bree accuses you * of "blundering," I have thought myself bound to send the enclosed letter † to 'Nature,' that is, if you in the least desire it. In this case please post it. If you do not *at all* wish it, I should rather prefer not sending it, and in this case please to tear it up. And I beg you to do the same, if you intend answering Dr. Bree yourself, as you will do it incomparably better than I should. Also please tear it up if you don't like the letter.

My dear Wallace, yours very sincerely,

CH. DARWIN.

* Mr. Wallace had reviewed Dr. Bree's book, 'An Exposition of Fallacies in the Hypothesis of Mr. Darwin,' in 'Nature,' July 25, 1872.

† "Bree on Darwinism." 'Nature,' Aug. 8, 1872. The letter is as follows :—"Permit me to state —though the statement is almost superfluous—that Mr. Wallace, in his review of Dr. Bree's work, gives with perfect correctness what I intended to express, and what I believe was expressed clearly, with respect to the probable position of man in the early part of his pedigree. As I have not seen Dr. Bree's recent work, and as his letter is unintelligible to me, I cannot even conjecture how he has so completely mistaken my meaning: but, perhaps, no one who has read Mr. Wallace's article, or who has read a work formerly published by Dr. Bree on the same subject as his recent one, will be surprised at any amount of misunderstanding on his part.—CHARLES DARWIN."

Aug. 3.

C. Darwin to A. R. Wallace.

Down, August 28, 1872.

MY DEAR WALLACE,—I have at last finished the gigantic job of reading Dr. Bastian's book,* and have been deeply interested by it. You wished to hear my impression, but it is not worth sending.

He seems to me an extremely able man, as, indeed, I thought when I read his first essay. His general argument in favour of Archebiosis † is wonderfully strong, though I cannot think much of some few of his arguments. The result is that I am bewildered and astonished by his statements, but am not convinced, though, on the whole, it seems to me probable that Archebiosis is true. I am not convinced, partly I think owing to the deductive cast of much of his reasoning; and I know not why, but I never feel convinced by deduction, even in the case of H. Spencer's writings. If Dr. Bastian's book had been turned upside down, and he had begun with the various cases of Heterogenesis, and then gone on to organic, and afterwards to saline solutions, and had then given his general arguments, I should have been, I believe, much more influenced. I suspect, however, that my chief difficulty is the effect of old convictions being stereotyped on my brain. I must have more evidence that germs, or the minutest fragments of the lowest forms, are always killed by 212° of Fahr. Perhaps the mere reiteration of the statements given by Dr. Bastian [of] other men, whose judgment I respect, and who have worked long on the lower organisms, would suffice to convince me. Here is a fine confession of intellectual weakness; but what an inexplicable frame of mind is that of belief!

As for Rotifers and Tardigrades being spontaneously gener-

* 'The Beginnings of Life.' H. C. Bastian, 1872.

† That is to say, Spontaneous Generation. For the distinction between Archebiosis and Heterogenesis, see Bastian, chapter vi.

ated, my mind can no more digest such statements, whether true or false, than my stomach can digest a lump of lead. Dr. Bastian is always comparing Archebiosis, as well as growth, to crystallisation ; but, on this view, a Rotifer or Tardigrade is adapted to its humble conditions of life by a happy accident, and this I cannot believe. . . . He must have worked with very impure materials in some cases, as plenty of organisms appeared in a saline solution not containing an atom of nitrogen.

I wholly disagree with Dr. Bastian about many points in his latter chapters. Thus the frequency of generalised forms in the older strata seems to me clearly to indicate the common descent with divergence of more recent forms. Notwithstanding all his sneers, I do not strike my colours as yet about Pangenesis. I should like to live to see Archebiosis proved true, for it would be a discovery of transcendent importance; or, if false, I should like to see it disproved, and the facts otherwise explained; but I shall not live to see all this. If ever proved, Dr. Bastian will have taken a prominent part in the work. How grand is the onward rush of science; it is enough to console us for the many errors which we have committed, and for our efforts being overlaid and forgotten in the mass of new facts and new views which are daily turning up.

This is all I have to say about Dr. Bastian's book, and it certainly has not been worth saying. . . .

C. Darwin to A. De Candolle.

Down, December 11, 1872.

MY DEAR SIR—I began reading your new book * sooner than I intended, and when I once began, I could not stop; and now you must allow me to thank you for the very great pleasure which it has given me. I have hardly ever read

* 'Histoire des Sciences et des Savants,' 1873.

anything more original and interesting than your treatment of the causes which favour the development of scientific men. The whole was quite new to me, and most curious. When I began your essay I was afraid that you were going to attack the principle of inheritance in relation to mind, but I soon found myself fully content to follow you and accept your limitations. I have felt, of course, special interest in the latter part of your work, but there was here less novelty to me. In many parts you do me much honour, and everywhere more than justice. Authors generally like to hear what points most strike different readers, so I will mention that of your shorter essays, that on the future prevalence of languages, and on vaccination interested me the most, as, indeed, did that on statistics, and free will. Great liability to certain diseases, being probably liable to atavism, is quite a new idea to me. At p. 322 you suggest that a young swallow ought to be separated, and then let loose in order to test the power of instinct; but nature annually performs this experiment, as old cuckoos migrate in England some weeks before the young birds of the same year. By the way, I have just used the forbidden word "nature," which, after reading your essay, I almost determined never to use again. There are very few remarks in your book to which I demur, but when you back up Asa Gray in saying that all instincts are congenital habits, I must protest.

Finally, will you permit me to ask you a question: have you yourself, or [has] some one who can be quite trusted, observed (p. 322) that the butterflies on the Alps are tamer than those on the lowlands? Do they belong to the same species? Has this fact been observed with more than one species? Are they brightly coloured kinds? I am especially curious about their alighting on the brightly coloured parts of ladies' dresses, more especially because I have been more than once assured that butterflies like bright colours, for instance, in India the scarlet leaves of Pointsettia.

Once again allow me to thank you for having sent me your work, and for the very unusual amount of pleasure which I have received in reading it.

With much respect, I remain, my dear Sir,

Yours very sincerely,

CHARLES DARWIN.

[The last revise of the 'Expression of the Emotions' was finished on August 22nd, 1872, and he wrote in his Diary:—"Has taken me about twelve months." As usual he had no belief in the possibility of the book being generally successful. The following passage in a letter to Haeckel serves to show that he had felt the writing of this book as a somewhat severe strain:—

"I have finished my little book on 'Expression,' and when it is published in November I will of course send you a copy, in case you would like to read it for amusement. I have resumed some old botanical work, and perhaps I shall never again attempt to discuss theoretical views.

"I am growing old and weak, and no man can tell when his intellectual powers begin to fail. Long life and happiness to you for your own sake and for that of science."

It was published in the autumn. The edition consisted of 7000, and of these 5267 copies were sold at Mr. Murray's sale in November. Two thousand were printed at the end of the year, and this proved a misfortune, as they did not afterwards sell so rapidly, and thus a mass of notes collected by the author was never employed for a second edition during his lifetime.

Among the reviews of the 'Expression of the Emotions' may be mentioned the not unfavourable notices in the *Athenæum*, Nov. 9, 1872, and the *Times*, Dec. 13, 1872. A good review by Mr. Wallace appeared in the 'Quarterly Journal of Science,' Jan. 1873. Mr. Wallace truly remarks that the book exhibits certain "characteristics of the author's mind in an eminent degree," namely, "the insatiable longing to discover the causes of the varied and complex phenomena presented by living things." He adds that in the case of the author "the restless curiosity of the child to know the 'what for?' the 'why?' and the 'how?' of everything" seems "never to have abated its force."

A writer in one of the theological reviews describes the book as "the most powerful and insidious" of all the author's works.

Professor Alexander Bain criticised the book in a postscript to the 'Senses and the Intellect;' to this essay the following letter refers:]

C. Darwin to Alexander Bain.

Down, October 9, 1873.

MY DEAR SIR,—I am particularly obliged to you for having sent me your essay. Your criticisms are all written in a quite fair spirit, and indeed no one who knows you or your works would expect anything else. What you say about the vagueness of what I have called the direct action of the nervous system, is perfectly just. I felt it so at the time, and even more of late. I confess that I have never been able fully to grasp your principle of spontaneity,* as well as some other of your points, so as to apply them to special cases,

* Professor Bain expounded his theory of Spontaneity in the essay here alluded to. It would be impossible to do justice to it within the limits of a foot-note. The following quotations may give some notion of it:—

"By Spontaneity I understand the readiness to pass into movement, in the absence of all stimulation whatever; the essential requisite being that the nerve-centres and muscles shall be fresh and vigorous. . . . . The gesticulations and the carols of young and active animals are mere overflow of nervous energy; and although they are very apt to concur with pleasing emotion, they have an independent source. . . . . They are not properly movements of expression; they express nothing at all except an abundant stock of physical power."

But as we look at everything from different points of view, it is not likely that we should agree closely.

I have been greatly pleased by what you say about the crying expression and about blushing. Did you read a review in a late 'Edinburgh'?* It was magnificently contemptuous towards myself and many others.

I retain a very pleasant recollection of our sojourn together at that delightful place, Moor Park.

With my renewed thanks, I remain, my dear Sir,

Yours sincerely,

CH. DARWIN.

C. Darwin to Mrs. Haliburton.†

Down, November 1 [1872].

MY DEAR MRS. HALIBURTON,—I dare say you will be surprised to hear from me. My object in writing now is to

* The review on the 'Expression of the Emotions' appeared in the April number of the 'Edinburgh Review,' 1873. The opening sentence is a fair sample of the general tone of the article: "Mr. Darwin has added another volume of amusing stories and grotesque illustrations to the remarkable series of works already devoted to the exposition and defence of the evolutionary hypothesis." A few other quotations may be worth giving. "His one-sided devotion to an *à priori* scheme of interpretation seems thus steadily tending to impair the author's hitherto unrivalled powers as an observer. However this may be, most impartial critics will, we think, admit that there is a marked falling off, both in philosophical tone and scientific interest, in the works produced since Mr. Darwin committed himself to the crude metaphysical conception so largely associated with his name." The article is directed against Evolution as a whole, almost as much as against the doctrines of the book under discussion. We find throughout plenty of that effective style of criticism which consists in the use of such expressions as "dogmatism," "intolerance," "presumptuous," "arrogant;" together with accusations of such various faults as "virtual abandonment of the inductive method," and the use of slang and vulgarisms.

The part of the article which seems to have interested my father is the discussion on the use which he ought to have made of painting and sculpture.

† Mrs. Haliburton is a daughter of my father's old friend, Mr. Owen of Woodhouse. Her husband, Judge Haliburton, was the well-known author of 'Sam Slick.'

say that I have just published a book on the 'Expression of the Emotions in Man and Animals;' and it has occurred to me that you might possibly like to read some parts of it; and I can hardly think that this would have been the case with any of the books which I have already published. So I send by this post my present book. Although I have had no communication with you or the other members of your family for so long a time, no scenes in my whole life pass so frequently or so vividly before my mind as those which relate to happy old days spent at Woodhouse. I should very much like to hear a little news about yourself and the other members of your family, if you will take the trouble to write to me. Formerly I used to glean some news about you from my sisters.

I have had many years of bad health and have not been able to visit anywhere; and now I feel very old. As long as I pass a perfectly uniform life, I am able to do some daily work in Natural History, which is still my passion, as it was in old days, when you used to laugh at me for collecting beetles with such zeal at Woodhouse. Excepting from my continued ill-health, which has excluded me from society, my life has been a very happy one; the greatest drawback being that several of my children have inherited from me feeble health. I hope with all my heart that you retain, at least to a large extent, the famous "Owen constitution." With sincere feelings of gratitude and affection for all bearing the name of Owen, I venture to sign myself,

Yours affectionately,

CHARLES DARWIN.

C. Darwin to Mrs. Haliburton.

Down, November 6 [1872].

MY DEAR SARAH,—I have been very much pleased by your letter, which I must call charming. I hardly ventured to think that you would have retained a friendly recollection of me for so many years. Yet I ought to have felt assured that you would remain as warm-hearted and as true-hearted as you have ever been from my earliest recollection. I know well how many grievous sorrows you have gone through; but I am very sorry to hear that your health is not good. In the spring or summer, when the weather is better, if you can summon up courage to pay us a visit here, both my wife, as she desires me to say, and myself, would be truly glad to see you, and I know that you would not care about being rather dull here. It would be a real pleasure to me to see you.—Thank you much for telling about your family,—much of which was new to me. How kind you all were to me as a boy, and you especially, and how much happiness I owe to you.

Believe me your affectionate and obliged friend,

CHARLES DARWIN.

P.S.—Perhaps you would like to see a photograph of me now that I am old.

1873.

[The only work (other than botanical) of this year was the preparation of a second edition of the 'Descent of Man,' the publication of which is referred to in the following chapter. This work was undertaken much against the grain, as he was at the time deeply immersed in the manuscript of 'Insectivorous Plants.' Thus he wrote to Mr. Wallace (November 19), "I never in my lifetime regretted an interruption so much as this new edition of the 'Descent.'" And later (in December) he wrote to Mr. Huxley: "The new edition of the 'Descent' has turned out an awful job. It took me ten days merely to glance over letters and reviews with criticisms and new facts. It is a devil of a job."

The work was continued until April 1, 1874, when he was

able to return to his much loved Drosera. He wrote to Mr. Murray:—

"I have at last finished, after above three months as hard work as I have ever had in my life, a corrected edition of the 'Descent,' and I much wish to have it printed off as soon as possible. As it is to be stereotyped I shall never touch it again."

The first of the miscellaneous letters of 1873 refers to a pleasant visit received from Colonel Higginson of Newport, U.S.]

C. Darwin to Thos. Wentworth Higginson.

Down, February 27th [1873].

MY DEAR SIR,—My wife has just finished reading aloud your 'Life with a Black Regiment,' and you must allow me to thank you heartily for the very great pleasure which it has in many ways given us. I always thought well of the negroes, from the little which I have seen of them; and I have been delighted to have my vague impressions confirmed, and their character and mental powers so ably discussed. When you were here I did not know of the noble position which you had filled. I had formerly read about the black regiments, but failed to connect your name with your admirable undertaking. Although we enjoyed greatly your visit to Down, my wife and myself have over and over again regretted that we did not know about the black regiment, as we should have greatly liked to have heard a little about the South from your own lips.

Your descriptions have vividly recalled walks taken forty years ago in Brazil. We have your collected Essays, which were kindly sent us by Mr. [Moncure] Conway, but have not yet had time to read them. I occasionally glean a little news of you in the 'Index'; and within the last hour have read an interesting article of yours on the progress of Free Thought.

Believe me, my dear Sir, with sincere admiration,

Yours very faithfully,

CH. DARWIN.

[On May 28th he sent the following answers to the questions that Mr. Galton was at that time addressing to various scientific men, in the course of the inquiry which is given in his 'English Men of Science, their Nature and Nurture,' 1874. With regard to the questions, my father wrote, "I have filled up the answers as well as I could, but it is simply impossible for me to estimate the degrees." For the sake of convenience, the questions and answers relating to "Nurture" are made to precede those on "Nature."

	Question	Answer
Education?	How taught?	I consider that all I have learnt of any value has been self-taught.
	Conducive to or restrictive of habits of observation.	Restrictive of observation, being almost entirely classical.
	Conducive to health or otherwise?	Yes.
	Peculiar merits?	None whatever.
	Chief omissions.	No mathematics or modern languages, nor any habits of observation or reasoning.
	Has the religious creed taught in your youth had any deterrent effect on the freedom of your researches?	No.
	Do your scientific tastes appear to have been innate?	Certainly innate.
	Were they determined by any and what events?	My innate taste for natural history strongly confirmed and directed by the voyage in the *Beagle*.

QUESTION.	YOURSELF.	YOUR FATHER.
Specify any interests that have been very actively pursued	Science, and field sports to a passionate degree during youth.	
Religion ?	Nominally to Church of England.	Nominally to Church of England.
Politics ?	Liberal or Radical.	Liberal.
Health ?	Good when young—bad for last 33 years.	Good throughout life, except from gout.
Height, &c. ?	Height ? 6 ft. Figure, &c. ? Spare, whilst young rather stout. Measurement round inside of hat. $22\frac{1}{4}$ in.	Height ? 6 ft. 2 in. Figure, &c. ? Very broad and corpulent.
	Colour of Hair ? Brown. Complexion ? Rather sallow.*	Colour of Hair ? Brown. Complexion ? Ruddy.
Temperament ?	Somewhat nervous.	Sanguine.
Energy of body, &c. ?	Energy shown by much activity, and whilst I had health, power of resisting fatigue. I and one other man were alone able to fetch water for a large party of officers and sailors utterly prostrated. Some of my expeditions in S. America were adventurous. An early riser in the morning.	Great power of endurance although feeling much fatigue, as after consultations, after long journeys ; very active—not restless—very early riser, no travels. My father said his father suffered much from sense of fatigue, that he worked very hard.

Energy of mind, &c.?	Shown by rigorous and long-continued work on same subject, as 20 years on the 'Origin of Species,' and 9 years on 'Cirripedia.'	Habitually very active mind—shown in conversation with a succession of people during the whole day.
Memory?	Memory very bad for dates, and for learning by rote; but good in retaining a general or vague recollection of many facts.	Wonderful memory for dates. In old age he told a person, reading aloud to him a book only read in youth, the passages which were coming—knew the birthdays and death, &c., of all friends and acquaintances.
Studiousness?	Very studious, but not large acquirements.	Not very studious or mentally receptive, except for facts in conversation—great collector of anecdotes.
Independence of Judgment?	I think fairly independent; but I can give no instances. I gave up common religious belief almost independently from my own reflections.	Free thinker in religious matters. Liberal, with rather a tendency to Toryism.
Originality, or Eccentricity?	—— thinks this applies to me; I do not think so—*i.e.*, as far as eccentricity. I suppose that I have shown originality in science, as I have made discoveries with regard to common objects.	Original character, had great personal influence, and power of producing fear of himself in others. He kept his accounts with great care in a peculiar way, in a number of separate little books, without any general ledger.
Special talents?	None, except for business as evinced by keeping accounts, replies to correspondence, and investing money very well. Very methodical in all my habits.	Practical business—made a large fortune and incurred no losses.
Strongly marked mental peculiarities, bearing on scientific success, and not specified above?	Steadiness—great curiosity about facts and their meaning. Some love of the new and marvellous.	Strong social affection and great sympathy in the pleasures of others. Sceptical as to new things. Curious as to facts. Great foresight. Not much public spirit—great generosity in giving money and assistance.

* His complexion was ruddy rather than sallow.

N 2

The following refers *inter alia* to a letter which appeared in 'Nature' (Sept. 25, 1873), "On the Males and Complemental Males of certain Cirripedes, and on Rudimentary Organs:"]

C. Darwin to E. Haeckel.

Down, September 25, 1873.

MY DEAR HÄCKEL,—I thank you for the present of your book,* and I am heartily glad to see its great success. You will do a wonderful amount of good in spreading the doctrine of Evolution, supporting it as you do by so many original observations. I have read the new preface with very great interest. The delay in the appearance of the English translation vexes and surprises me, for I have never been able to read it thoroughly in German, and I shall assuredly do so when it appears in English. Has the problem of the later stages of reduction of useless structures ever perplexed you? This problem has of late caused me much perplexity. I have just written a letter to 'Nature' with a hypothetical explanation of this difficulty, and I will send you the paper with the passage marked. I will at the same time send a paper which has interested me; it need not be returned. It contains a singular statement bearing on so-called Spontaneous Generation. I much wish that this latter question could be settled, but I see no prospect of it. If it could be proved true this would be most important to us. . . .

Wishing you every success in your admirable labours,
I remain, my dear Häckel, yours very sincerely,
CHARLES DARWIN.

* 'Schöpfungs-Geschichte,' 4th ed. The translation ('The History of Creation') was not published until 1876.

CHAPTER V.

MISCELLANEA, INCLUDING SECOND EDITIONS OF 'CORAL REEFS,' THE 'DESCENT OF MAN,' AND THE 'VARIATION OF ANIMALS AND PLANTS.'

1874 AND 1875.

[THE year 1874 was given up to 'Insectivorous Plants,' with the exception of the months devoted to the second edition of the 'Descent of Man,' (see Vol. III. p. 175) and with the further exception of the time given to a second edition of his 'Coral Reefs' (1874). The Preface to the latter states that new facts have been added, the whole book revised, and "the latter chapters almost rewritten." In the Appendix some account is given of Professor Semper's objections, and this was the occasion of correspondence between that naturalist and my father. In Professor Semper's volume, 'Animal Life' (one of the International Series), the author calls attention to the subject in the following passage which I give in German, the published English translation being, as it seems to me, incorrect: "Es scheint mir als ob er in der zweiten Ausgabe seines allgemein bekannten Werks über Korallenriffe einem Irrthume über meine Beobachtungen zum Opfer gefallen ist, indem er die Angaben, die ich allerdings bisher immer nur sehr kurz gehalten hatte, vollständig falsch wiedergegeben hat."

The proof-sheets containing this passage were sent by Professor Semper to my father before 'Animal Life' was published, and this was the occasion for the following letter, which was afterwards published in Professor Semper's book.]

C. Darwin to K. Semper.

Down, October 2, 1879.

My dear Professor Semper,—I thank you for your extremely kind letter of the 19th, and for the proof-sheets. I believe that I understand all, excepting one or two sentences, where my imperfect knowledge of German has interfered. This is my sole and poor excuse for the mistake which I made in the second edition of my 'Coral' book. Your account of the Pellew Islands is a fine addition to our knowledge on coral reefs. I have very little to say on the subject, even if I had formerly read your account and seen your maps, but had known nothing of the proofs of recent elevation, and of your belief that the islands have not since subsided. I have no doubt that I should have considered them as formed during subsidence. But I should have been much troubled in my mind by the sea not being so deep as it usually is round atolls, and by the reef on one side sloping so gradually beneath the sea; for this latter fact, as far as my memory serves me, is a very unusual and almost unparalleled case. I always foresaw that a bank at the proper depth beneath the surface would give rise to a reef which could not be distinguished from an atoll, formed during subsidence. I must still adhere to my opinion, that the atolls and barrier reefs in the middle of the Pacific and Indian Oceans indicate subsidence; but I fully agree with you that such cases as that of the Pellew Islands, if of at all frequent occurrence, would make my general conclusions of very little value. Future observers must decide between us. It will be a strange fact if there has not been subsidence of the beds of the great oceans, and if this has not affected the forms of the coral reefs.

In the last three pages of the last sheet sent I am extremely glad to see that you are going to treat of the dispersion of animals. Your preliminary remarks seem to me quite excellent. There is nothing about M. Wagner, as I expected to find. I suppose that you have seen Moseley's last book, which contains some good observations on dispersion.

I am glad that your book will appear in English, for then I can read it with ease. Pray believe me,

Yours very sincerely,
CHARLES DARWIN.

[The most recent criticism on the Coral-reef theory is by Mr. Murray, one of the staff of the *Challenger*, who read a paper before the Royal Society of Edinburgh, April 5, 1880.* The chief point brought forward is the possibility of the building up of submarine mountains, which may serve as foundations for coral reefs. Mr. Murray also seeks to prove that "the chief features of coral reefs and islands can be accounted for without calling in the aid of great and general subsidence." The following letter refers to this subject :]

C. Darwin to A. Agassiz.

Down, May 5, 1881.

. . . You will have seen Mr. Murray's views on the formation of atolls and barrier reefs. Before publishing my book, I thought long over the same view, but only as far as ordinary marine organisms are concerned, for at that time little was known of the multitude of minute oceanic organisms. I rejected this view, as from the few dredgings made in the *Beagle*, in the south temperate regions, I concluded that shells, the smaller corals, &c., decayed, and were dissolved, when not protected by the deposition of sediment, and sediment could not accumulate in the open ocean. Certainly, shells, &c., were in several cases completely rotten, and crumbled into mud between my fingers; but you will know well whether

* An abstract is published in vol. x. of the 'Proceedings,' p. 505, and in 'Nature,' August 12, 1880.

this is in any degree common. I have expressly said that a bank at the proper depth would give rise to an atoll, which could not be distinguished from one formed during subsidence. I can, however, hardly believe in the former presence of as many banks (there having been no subsidence) as there are atolls in the great oceans, within a reasonable depth, on which minute oceanic organisms could have accumulated to the thickness of many hundred feet. . . . Pray forgive me for troubling you at such length, but it has occurred [to me] that you might be disposed to give, after your wide experience, your judgment. If I am wrong, the sooner I am knocked on the head and annihilated so much the better. It still seems to me a marvellous thing that there should not have been much, and long continued, subsidence in the beds of the great oceans. I wish that some doubly rich millionaire would take it into his head to have borings made in some of the Pacific and Indian atolls, and bring home cores for slicing from a depth of 500 or 600 feet. . . .

[The second edition of the 'Descent of Man' was published in the autumn of 1874. Some severe remarks on the "monistic hypothesis" appeared in the July * number of the 'Quarterly Review' (p. 45). The reviewer expresses his astonishment at the ignorance of certain elementary distinctions and principles (*e.g.* with regard to the *verbum mentale*) exhibited, among others, by Mr. Darwin, who "does not exhibit the faintest indication of having grasped them, yet a clear perception of them, and a direct and detailed examination of his facts with regard to them, was a *sine quâ non* for attempting, with a chance of success, the solution of the mystery as to the descent of man."

Some further criticisms of a later date may be here alluded to. In the 'Academy,' 1876 (pp. 562, 587), appeared a review of Mr. Mivart's 'Lessons from Nature,' by Mr. Wallace.

* The review necessarily deals with the first edition of the 'Descent of Man.'

When considering the part of Mr. Mivart's book relating to Natural and Sexual Selection, Mr. Wallace says: "In his violent attack on Mr. Darwin's theories our author uses unusually strong language. Not content with mere argument, he expresses 'reprobation of Mr. Darwin's views'; and asserts that though he (Mr. Darwin) has been obliged, virtually, to give up his theory, it is still maintained by Darwinians with 'unscrupulous audacity,' and the actual repudiation of it concealed by the 'conspiracy of silence.'" Mr. Wallace goes on to show that these charges are without foundation, and points out that, "If there is one thing more than another for which Mr. Darwin is pre-eminent among modern literary and scientific men, it is for his perfect literary honesty, his self-abnegation in confessing himself wrong, and the eager haste with which he proclaims and even magnifies small errors in his works, for the most part discovered by himself."

The following extract from a letter to Mr. Wallace (June 17th) refers to Mr. Mivart's statement ('Lessons from Nature,' p. 144) that Mr. Darwin at first studiously disguised his views as to the "bestiality of man":—

"I have only just heard of and procured your two articles in the 'Academy.' I thank you most cordially for your generous defence of me against Mr. Mivart. In the 'Origin' I did not discuss the derivation of any one species; but that I might not be accused of concealing my opinion, I went out of my way, and inserted a sentence which seemed to me (and still so seems) to disclose plainly my belief. This was quoted in my 'Descent of Man.' Therefore it is very unjust . . . of Mr. Mivart to accuse me of base fraudulent concealment."

The letter which here follows is of interest in connection with the discussion, in the 'Descent of Man,' on the origin of the musical sense in man:]

*C. Darwin to E. Gurney.**

Down, July 8, 1876.

MY DEAR MR. GURNEY,—I have read your article † with much interest, except the latter part, which soared above my ken. I am greatly pleased that you uphold my views to a certain extent. Your criticism of the rasping noise made by insects being necessarily rhythmical is very good; but though not made intentionally, it may be pleasing to the females, from the nerve cells being nearly similar in function throughout the animal kingdom. With respect to your letter, I believe that I understand your meaning, and agree with you. I never supposed that the different degrees and kinds of pleasure derived from different music could be explained by the musical powers of our semi-human progenitors. Does not the fact that different people belonging to the same civilized nation are very differently affected by the same music, almost show that these diversities of taste and pleasure have been acquired during their individual lives? Your simile of architecture seems to me particularly good; for in this case the appreciation almost must be individual, though possibly the sense of sublimity excited by a grand cathedral may have some connection with the vague feelings of terror and superstition in our savage ancestors, when they entered a great cavern or gloomy forest. I wish some one could analyse the feeling of sublimity. It amuses me to think how horrified some high-flying æsthetic men will be, at your encouraging such low degraded views as mine.

Believe me, yours very sincerely,

CHARLES DARWIN.

[The letters which follow are of a miscellaneous interest. The first extract (from a letter, Jan. 18, 1874) refers to a spiritualistic séance, held at Erasmus Darwin's house, 6

* Author of 'The Power of Sound.'

† "Some disputed Points in Music."—'Fortnightly Review,' July 1876.

Queen Anne Street, under the auspices of a well-known medium:

"... We had grand fun, one afternoon, for George hired a medium, who made the chairs, a flute, a bell, and candlestick, and fiery points jump about in my brother's dining-room, in a manner that astounded every one, and took away all their breaths. It was in the dark, but George and Hensleigh Wedgwood held the medium's hands and feet on both sides all the time. I found it so hot and tiring that I went away before all these astounding miracles, or jugglery, took place. How the man could possibly do what was done passes my understanding. I came downstairs, and saw all the chairs, &c., on the table, which had been lifted over the heads of those sitting round it.

The Lord have mercy on us all, if we have to believe in such rubbish. F. Galton was there, and says it was a good séance. . . ."

The séance in question led to a smaller and more carefully organised one being undertaken, at which Mr. Huxley was present, and on which he reported to my father:]

C. Darwin to Professor T. H. Huxley.

Down, January 29 [1874].

MY DEAR HUXLEY,—It was very good of you to write so long an account. Though the séance did tire you so much it was, I think, really worth the exertion, as the same sort of things are done at all the séances, even at ——'s; and now to my mind an enormous weight of evidence would be requisite to make one believe in anything beyond mere trickery. . . . I am pleased to think that I declared to all my family, the day before yesterday, that the more I thought of all that I had heard happened at Queen Anne St., the more convinced I was it was all imposture . . . . my theory was that [the medium] managed to get the two men on each side of him to hold each other's hands, instead of his, and that he was thus free to perform his antics. I am very glad that I issued my ukase to you to attend.

Yours affectionately,
CH. DARWIN.

[In the spring of this year (1874) he read a book which gave him great pleasure and of which he often spoke with admiration:—The 'Naturalist in Nicaragua,' by the late Thomas Belt. Mr. Belt, whose untimely death may well be deplored by naturalists, was by profession an Engineer, so that all his admirable observations in natural history, in Nicaragua and elsewhere, were the fruit of his leisure. The book is direct and vivid in style and is full of description and suggestive discussions. With reference to it my father wrote to Sir J. D. Hooker:—

"Belt I have read, and I am delighted that you like it so much; it appears to me the best of all natural history journals which have ever been published."]

C. Darwin to the Marquis de Saporta.

Down, May 30, 1874.

DEAR SIR,—I have been very neglectful in not having sooner thanked you for your kindness in having sent me your 'Études sur la Végétation,' &c., and other memoirs. I have read several of them with very great interest, and nothing can be more important, in my opinion, than your evidence of the extremely slow and gradual manner in which specific forms change. I observe that M. A. De Candolle has lately quoted you on this head *versus* Heer. I hope that you may be able to throw light on the question whether such protean, or polymorphic forms, as those of Rubus, Hieracium, &c., at the present day, are those which generate new species; as for myself, I have always felt some doubt on this head. I trust that you may soon bring many of your countrymen to believe in Evolution, and my name will then perhaps cease to be scorned. With the most sincere respect, I remain, dear Sir,

Yours faithfully,

CH. DARWIN.

C. Darwin to Asa Gray.

Down, June 5 [1874].

MY DEAR GRAY,—I have now read your article * in 'Nature,' and the last two paragraphs were not included in the slip sent before. I wrote yesterday and cannot remember exactly what I said, and now cannot be easy without again telling you how profoundly I have been gratified. Every one, I suppose, occasionally thinks that he has worked in vain, and when one of these fits overtakes me, I will think of your article, and if that does not dispel the evil spirit, I shall know that I am at the time a little bit insane, as we all are occasionally.

What you say about Teleology† pleases me especially, and I do not think any one else ‡ has ever noticed the point. I have always said you were the man to hit the nail on the head.

Yours gratefully and affectionately,

CH. DARWIN.

[As a contribution to the history of the reception of the 'Origin of Species,' the meeting of the British Association in 1874, at Belfast, should be mentioned. It is memorable for

* The article, "Charles Darwin," in the series of *Scientific Worthies* ('Nature,' June 4, 1874). This admirable estimate of my father's work in science is given in the form of a comparison and contrast between Robert Brown and Charles Darwin.

† "Let us recognise Darwin's great service to Natural Science in bringing back to it Teleology: so that instead of Morphology *versus* Teleology, we shall have Morphology wedded to Teleology."

‡ Similar remarks had been previously made by Mr. Huxley. See Vol. II. p. 201.

Professor Tyndall's brilliant presidential address, in which a sketch of the history of Evolution is given, culminating in an eloquent analysis of the 'Origin of Species,' and of the nature of its great success. With regard to Prof. Tyndall's address, Lyell wrote ('Life,' vol. ii. p. 455) congratulating my father on the meeting, "on which occasion you and your theory of Evolution may be fairly said to have had an ovation." In the same letter Sir Charles speaks of a paper * by Professor Judd, and it is to this that the following letter refers :]

C. Darwin to C. Lyell.

Down, September 23, 1874.

MY DEAR LYELL,—I suppose that you have returned, or will soon return, to London ; † and, I hope, reinvigorated by your outing. In your last letter you spoke of Mr. Judd's paper on the Volcanoes of the Hebrides. I have just finished it, and to ease my mind must express my extreme admiration.

It is years since I have read a purely geological paper which has interested me so greatly. I was all the more interested, as in the Cordillera I often speculated on the sources of the deluges of submarine porphyritic lavas, of which they are built ; and, as I have stated, I saw to a certain extent the causes of the obliteration of the points of eruption. I was also not a little pleased to see my volcanic book quoted, for I thought it was completely dead and forgotten. What fine work will Mr. Judd assuredly do ! . . . . Now I have eased my mind ; and so farewell, with both E. D.'s and C. D.'s very kind remembrances to Miss Lyell.

Yours affectionately,

CHARLES DARWIN.

* "On the Ancient Volcanoes of the Highlands."—'Journal of Geolog. Soc.,' 1874.

† Sir Charles Lyell returned from Scotland towards the end of September.

[Sir Charles Lyell's reply to the above letter must have been one of the latest that my father received from his old friend, and it is with this letter that the last volume of Lyell's published correspondence closes.]

C. Darwin to Aug. Forel.

Down, October 15, 1874.

My dear Sir,—I have now read the whole of your admirable work * and seldom in my life have I been more interested by any book. There are so many interesting facts and discussions, that I hardly know which to specify; but I think, firstly, the newest points to me have been about the size of the brain in the three sexes, together with your suggestion that increase of mind-power may have led to the sterility of the workers. Secondly about the battles of the ants, and your curious account of the enraged ants being held by their comrades until they calmed down. Thirdly, the evidence of ants of the same community being the offspring of brothers and sisters. You admit, I think, that new communities will often be the product of a cross between not-related ants. Fritz Müller has made some interesting observations on this head with respect to Termites. The case of *Anergates* is most perplexing in many ways, but I have such faith in the law of occasional crossing that I believe an explanation will hereafter be found, such as the dimorphism of either sex and the occasional production of winged males. I see that you are puzzled how ants of the same community recognize each other; I once placed two (*F. rufa*) in a pill-box smelling strongly of asafœtida and after a day returned them to their homes; they were threatened, but at last recognized. I made the trial thinking that they might know each other by

* 'Les Fourmis de la Suisse,' 4to, 1874.

their odour; but this cannot have been the case, and I have often fancied that they must have some common signal. Your last chapter is one great mass of wonderful facts and suggestions, and the whole profoundly interesting. I have seldom been more gratified than by [your] honourable mention of my work.

I should like to tell you one little observation which I made with care many years ago; I saw ants (*Formica rufa*) carrying cocoons from a nest which was the largest I ever saw and which was well known to all the country people near, and an old man, apparently about eighty years of age, told me that he had known it ever since he was a boy. The ants carrying the cocoons did not appear to be emigrating; following the line, I saw many ascending a tall fir-tree still carrying their cocoons. But when I looked closely I found that all the cocoons were empty cases. This astonished me, and next day I got a man to observe with me, and we again saw ants bringing empty cocoons out of the nest; each of us fixed on one ant and slowly followed it, and repeated the observation on many others. We thus found that some ants soon dropped their empty cocoons; others carried them for many yards, as much as thirty paces, and others carried them high up the fir-tree out of sight. Now here I think we have one instinct in contest with another and mistaken one. The first instinct being to carry the empty cocoons out of the nest, and it would have been sufficient to have laid them on the heap of rubbish, as the first breath of wind would have blown them away. And then came in the contest with the other very powerful instinct of preserving and carrying their cocoons as long as possible; and this they could not help doing although the cocoons were empty. According as the one or other instinct was the stronger in each individual ant, so did it carry the empty cocoon to a greater or less distance. If this little observation should ever prove of any use to you, you are quite at liberty to use it. Again thanking you cordially for the great pleasure which your work has given me, I remain with much respect,

Yours sincerely,
CH. DARWIN.

P.S.—If you read English easily I should like to send you Mr. Belt's book, as I think you would like it as much as did Fritz Müller.

C. Darwin to J. Fiske.

Down, December 8, 1874.

MY DEAR SIR,—You must allow me to thank you for the very great interest with which I have at last slowly read the whole of your work.* I have long wished to know something about the views of the many great men whose doctrines you give. With the exception of special points I did not even understand H. Spencer's general doctrine; for his style is too hard work for me. I never in my life read so lucid an expositor (and therefore thinker) as you are; and I think that I understand nearly the whole—perhaps less clearly about Cosmic Theism and Causation than other parts. It is hopeless to attempt out of so much to specify what has interested me most, and probably you would not care to hear. I wish some chemist would attempt to ascertain the result of the cooling of heated gases of the proper kinds, in relation to your hypothesis of the origin of living matter. It pleased me to find that here and there I had arrived from my own crude thoughts at some of the same conclusions with you; though I could seldom or never have given my reasons for such conclusions. I find that my mind is so fixed by the inductive method, that I cannot appreciate deductive reasoning: I must begin with a good body of facts and not from a principle (in which I always suspect some fallacy) and then

* 'Outlines of Cosmic Philosophy,' 2 vols. 8vo. 1874.

as much deduction as you please. This may be very narrow-minded; but the result is that such parts of H. Spencer as I have read with care impress my mind with the idea of his inexhaustible wealth of suggestion, but never convince me; and so I find it with some others. I believe the cause to lie in the frequency with which I have found first-formed theories [to be] erroneous. I thank you for the honourable mention which you make of my works. Parts of the 'Descent of Man' must have appeared laughably weak to you: nevertheless, I have sent you a new edition just published. Thanking you for the profound interest and profit with which I have read your work, I remain,

My dear Sir, yours very faithfully,

CH. DARWIN.

1875.

[The only work, not purely botanical, which occupied my father in the present year was the correction of the second edition of 'The Variation of Animals and Plants,' and on this he was engaged from the beginning of July till October 3rd. The rest of the year was taken up with his work on insectivorous plants, and on cross-fertilisation, as will be shown in a later chapter. The chief alterations in the second edition of 'Animals and Plants' are in the eleventh chapter on "Bud-variation and on certain anomalous modes of reproduction;" the chapter on Pangenesis "was also largely altered and re-modelled." He mentions briefly some of the authors who have noticed the doctrine. Professor Delpino's 'Sulla Darwiniana Teoria della Pangenesi' (1869), an adverse but fair criticism, seems to have impressed him as valuable. Of another critic my father characteristically says,* "Dr. Lionel Beale ('Nature,' May 11, 1871, p. 26) sneers at the whole doctrine with much acerbity and some justice." He also

* 'Animals and Plants,' 2nd edit. vol. ii. p. 350.

points out that, in Mantegazza's 'Elementi di Igiene,' the theory of Pangenesis was clearly forestalled.

In connection with this subject, a letter of my father's to 'Nature' (April 27, 1871) should be mentioned. A paper by Mr. Galton had been read before the Royal Society (March 30, 1871) in which were described experiments, on intertransfusion of blood, designed to test the truth of the hypothesis of pangenesis. My father, while giving all due credit to Mr. Galton for his ingenious experiments, does not allow that pangenesis has "as yet received its death-blow, though from presenting so many vulnerable points its life is always in jeopardy."

He seems to have found the work of correcting very wearisome, for he wrote:—

"I have no news about myself, as I am merely slaving over the sickening work of preparing new editions. I wish I could get a touch of poor Lyell's feelings, that it was delightful to improve a sentence, like a painter improving a picture."

The feeling of effort or strain over this piece of work, is shown in a letter to Professor Haeckel:—

"What I shall do in future if I live, Heaven only knows; I ought perhaps to avoid general and large subjects, as too difficult for me with my advancing years, and I suppose enfeebled brain."

At the end of March, in this year, the portrait for which he was sitting to Mr. Ouless was finished. He felt the sittings a great fatigue, in spite of Mr. Ouless's considerate desire to spare him as far as was possible. In a letter to Sir J. D. Hooker he wrote, "I look a very venerable, acute, melancholy old dog; whether I really look so I do not know." The picture is in the possession of the family, and is known to many through M. Rajon's etching. Mr. Ouless's portrait is, in my opinion, the finest representation of my father that has been produced.

The following letter refers to the death of Sir Charles Lyell, which took place on February 22nd, 1875, in his seventy-eighth year.]

*C. Darwin to Miss Buckley (now Mrs. Fisher).**

Down, February 23, 1875.

MY DEAR MISS BUCKLEY,—I am grieved to hear of the death of my old and kind friend, though I knew that it could not be long delayed, and that it was a happy thing that his life should not have been prolonged, as I suppose that his mind would inevitably have suffered. I am glad that Lady Lyell † has been saved this terrible blow. His death makes me think of the time when I first saw him, and how full of sympathy and interest he was about what I could tell him of coral reefs and South America. I think that this sympathy with the work of every other naturalist was one of the finest features of his character. How completely he revolutionised Geology: for I can remember something of pre-Lyellian days.

I never forget that almost everything which I have done in science I owe to the study of his great works. Well, he has had a grand and happy career, and no one ever worked with a truer zeal in a noble cause. It seems strange to me that I shall never again sit with him and Lady Lyell at their breakfast. I am very much obliged to you for having so kindly written to me.

Pray give our kindest remembrances to Miss Lyell, and I hope that she has not suffered much in health, from fatigue and anxiety.

Believe me, my dear Miss Buckley,

Yours very sincerely,

CHARLES DARWIN.

* Mrs. Fisher acted as Secretary to Sir Charles Lyell.

† Lady Lyell died in 1873.

C. Darwin to J. D. Hooker.

Down, February 25 [1875].

MY DEAR HOOKER,—Your letter so full of feeling has interested me greatly. I cannot say that I felt his [Lyell's] death much, for I fully expected it, and have looked for some little time at his career as finished.

I dreaded nothing so much as his surviving with impaired mental powers. He was, indeed, a noble man in very many ways; perhaps in none more than in his warm sympathy with the work of others. How vividly I can recall my first conversation with him, and how he astonished me by his interest in what I told him. How grand also was his candour and pure love of truth. Well, he is gone, and I feel as if we were all soon to go. . . . I am deeply rejoiced about Westminster Abbey,* the possibility of which had not occurred to me when I wrote before. I did think that his works were the most enduring of all testimonials (as you say) to him; but then I did not like the idea of his passing away with no outward sign of what scientific men thought of his merits. Now all this is changed, and nothing can be better than Westminster Abbey. Mrs. Lyell has asked me to be one of the pall-bearers, but I have written to say that I dared not, as I should so likely fail in the midst of the ceremony, and have my head whirling off my shoulders. All this affair must have cost you much fatigue and worry, and how I do wish you were out of England. . . .

[In 1881 he wrote to Mrs. Fisher in reference to her article on Sir Charles Lyell in the 'Encyclopædia Britannica':—

"For such a publication I suppose you do not want to say much about his private character, otherwise his strong sense of humour and love of society might have been added. Also his extreme interest in the progress of the world, and in the

* Sir Charles Lyell was buried in Westminster Abbey.

happiness of mankind. Also his freedom from all religious bigotry, though these perhaps would be a superfluity."

The following refers to the Zoological station at Naples, a subject on which my father felt an enthusiastic interest :]

C. Darwin to Anton Dohrn.

Down [1875 ?].

MY DEAR DR. DOHRN,—Many thanks for your most kind letter. I most heartily rejoice at your improved health and at the success of your grand undertaking, which will have so much influence on the progress of Zoology throughout Europe.

If we look to England alone, what capital work has already been done at the Station by Balfour and Ray Lankester..... When you come to England, I suppose that you will bring Mrs. Dohrn, and we shall be delighted to see you both here. I have often boasted that I have had a live Uhlan in my house! It will be very interesting to me to read your new views on the ancestry of the Vertebrates. I shall be sorry to give up the Ascidians, to whom I feel profound gratitude ; but the great thing, as it appears to me, is that any link whatever should be found between the main divisions of the Animal Kingdom. . . .

C. Darwin to August Weismann.

Down, December 6, 1875.

MY DEAR SIR,—I have been profoundly interested by your essay on Amblystoma,* and think that you have removed a great stumbling-block in the way of Evolution. I once thought of reversion in this case ; but in a crude and imperfect manner. I write now to call your attention to the sterility of moths when hatched out of their proper season ; I give references in chapter 18 of my 'Variation under Domestication' (vol. ii.

* 'Umwandlung des Axolotl.'

p. 157, of English edition), and these cases illustrate, I think, the sterility of Amblystoma. Would it not be worth while to examine the reproductive organs of those individuals of *wingless* Hemiptera which occasionally have wings, as in the case of the bed-bug? I think I have heard that the females of Mutilla sometimes have wings. These cases must be due to reversion. I dare say many anomalous cases will be hereafter explained on the same principle.

I hinted at this explanation in the extraordinary case of the black-shouldered peacock, the so-called *Pavo nigripennis* given in my 'Var. under Domest.;' and I might have been bolder, as the variety is in many respects intermediate between the two known species.

With much respect,

Yours sincerely,

CH. DARWIN.

THE VIVISECTION QUESTION.

[It was in November 1875 that my father gave his evidence before the Royal Commission on Vivisection.* I have, therefore, placed together here the matter relating to this subject, irrespective of date. Something has already been said of my father's strong feeling with regard to suffering † both in man and beast. It was indeed one of the strongest feelings in his nature, and was exemplified in matters small and great, in his sympathy with the educational miseries of dancing dogs, or in his horror at the sufferings of slaves.

* See Vol. I. p. 141.

† He once made an attempt to free a patient in a mad-house, who (as he wrongly supposed) was sane. He had some correspondence with the gardener at the asylum, and on one occasion he found a letter from a patient enclosed with one from the gardener. The letter was rational in tone and declared that the writer was sane and wrongfully confined.

My father wrote to the Lunacy Commissioners (without explaining the source of his information) and in due time heard that the man had been visited by the Commissioners, and that he was certainly insane.

[Some

The remembrance of screams, or other sounds heard in Brazil, when he was powerless to interfere with what he believed to be the torture of a slave, haunted him for years, especially at night. In smaller matters, where he could interfere, he did so vigorously. He returned one day from his walk pale and faint from having seen a horse ill-used, and from the agitation of violently remonstrating with the man. On another occasion he saw a horse-breaker teaching his son to ride, the little boy was frightened and the man was rough; my father stopped, and jumping out of the carriage reproved the man in no measured terms.

One other little incident may be mentioned, showing that his humanity to animals was well known in his own neighbourhood. A visitor, driving from Orpington to Down, told the cabman to go faster. "Why," said the man, "if I had whipped the horse *this* much, driving Mr. Darwin, he would have got out of the carriage and abused me well."

With respect to the special point under consideration,—the sufferings of animals subjected to experiment,—nothing could show a stronger feeling than the following extract from a letter to Professor Ray Lankester (March 22, 1871):—

"You ask about my opinion on vivisection. I quite agree that it is justifiable for real investigations on physiology; but not for mere damnable and detestable curiosity. It is a subject which makes me sick with horror, so I will not say another word about it, else I shall not sleep to-night."

An extract from Sir Thomas Farrer's notes shows how strongly he expressed himself in a similar manner in conversation:—

"The last time I had any conversation with him was at my house in Bryanston Square, just before one of his last seizures. He was then deeply interested in the vivisection question;

Some time afterwards the patient was discharged, and wrote to thank my father for his interference, adding that he had undoubtedly been insane when he wrote his former letter.

and what he said made a deep impression on me. He was a man eminently fond of animals and tender to them; he would not knowingly have inflicted pain on a living creature; but he entertained the strongest opinion that to prohibit experiments on living animals, would be to put a stop to the knowledge of and the remedies for pain and disease."

The Anti-Vivisection agitation, to which the following letters refer, seems to have become specially active in 1874, as may be seen, *e.g.* by the index to 'Nature' for that year, in which the word "Vivisection" suddenly comes into prominence. But before that date the subject had received the earnest attention of biologists. Thus at the Liverpool Meeting of the British Association in 1870, a Committee was appointed, whose report defined the circumstances and conditions under which, in the opinion of the signatories, experiments on living animals were justifiable. In the spring of 1875, Lord Hartismere introduced a Bill into the Upper House to regulate the course of physiological research. Shortly afterwards a Bill more just towards science in its provisions was introduced to the House of Commons by Messrs. Lyon Playfair, Walpole, and Ashley. It was however, withdrawn on the appointment of a Royal Commission to inquire into the whole question. The Commissioners were Lords Cardwell and Winmarleigh, Mr. W. E. Forster, Sir J. B. Karslake, Mr. Huxley, Professor Erichssen, and Mr. R. H. Hutton: they commenced their inquiry in July, 1875, and the Report was published early in the following year.

In the early summer of 1876, Lord Carnarvon's Bill, entitled, "An Act to amend the Law relating to Cruelty to Animals," was introduced. It cannot be denied that the framers of this Bill, yielding to the unreasonable clamour of the public, went far beyond the recommendations of the Royal Commission. As a correspondent in 'Nature' put it (1876, p. 248), "the evidence on the strength of which legislation was recommended went beyond the facts, the Report went beyond the evidence, the Recommendations beyond the Report; and the Bill can hardly be said to have gone beyond the Recommendations; but rather to have contradicted them."

The legislation which my father worked for, as described in the following letters, was practically what was introduced as Dr. Lyon Playfair's Bill.]

*C. Darwin to Mrs. Litchfield.**

January 4, 1875.

MY DEAR H.—Your letter has led me to think over vivisection (I wish some new word like anæs-section could be invented †) for some hours, and I will jot down my conclusions, which will appear very unsatisfactory to you. I have long thought physiology one of the greatest of sciences, sure sooner, or more probably later, greatly to benefit mankind; but, judging from all other sciences, the benefits will accrue only indirectly in the search for abstract truth. It is certain that physiology can progress only by experiments on living animals. Therefore the proposal to limit research to points of which we can now see the bearings in regard to health, &c., I look at as puerile. I thought at first it would be good to limit vivisection to public laboratories; but I have heard only of those in London and Cambridge, and I think Oxford; but probably there may be a few others. Therefore only men living in a few great towns would carry on investigation, and this I should consider a great evil. If private men were permitted to work in their own houses, and required a licence, I do not see who is to determine whether any particular man should receive one. It is young unknown men who are the

* His daughter.

† He communicated to 'Nature' (Sept. 30, 1880) an article by Dr. Wilder, of Cornell University, an abstract of which was published (p. 517). Dr. Wilder advocated the use of the word 'Callisection' for painless operations on animals.

most likely to do good work. I would gladly punish severely any one who operated on an animal not rendered insensible, if the experiment made this possible; but here again I do not see that a magistrate or jury could possibly determine such a point. Therefore I conclude, if (as is likely) some experiments have been tried too often, or anæsthetics have not been used when they could have been, the cure must be in the improvement of humanitarian feelings. Under this point of view I have rejoiced at the present agitation. If stringent laws are passed, and this is likely, seeing how unscientific the House of Commons is, and that the gentlemen of England are humane, as long as their sports are not considered, which entail a hundred or thousand-fold more suffering than the experiments of physiologists—if such laws are passed, the result will assuredly be that physiology, which has been until within the last few years at a standstill in England, will languish or quite cease. It will then be carried on solely on the Continent; and there will be so many the fewer workers on this grand subject, and this I should greatly regret. By the way, F. Balfour, who has worked for two or three years in the laboratory at Cambridge, declares to George that he has never seen an experiment, except with animals rendered insensible. No doubt the names of doctors will have great weight with the House of Commons; but very many practitioners neither know nor care anything about the progress of knowledge. I cannot at present see my way to sign any petition, without hearing what physiologists thought would be its effect, and then judging for myself. I certainly could not sign the paper sent me by Miss Cobbe, with its monstrous (as it seems to me) attack on Virchow for experimenting on the Trichinæ. I am tired and so no more.

Yours affectionately,

CHARLES DARWIN.

C. Darwin to J. D. Hooker.

Down, April 14 [1875].

MY DEAR HOOKER.—I worked all the time in London on the vivisection question; and we now think it advisable to go further than a mere petition. Litchfield* drew up a sketch of a Bill, the essential features of which have been approved by Sanderson, Simon and Huxley, and from conversation, will, I believe, be approved by Paget, and almost certainly, I think, by Michael Foster. Sanderson, Simon and Paget wish me to see Lord Derby, and endeavour to gain his advocacy with the Home Secretary. Now, if this is carried into effect, it will be of great importance to me to be able to say that the Bill in its essential features has the approval of some half-dozen eminent scientific men. I have therefore asked Litchfield to enclose a copy to you in its first rough form; and if it is not essentially modified, may I say that it meets with your approval as President of the Royal Society? The object is to protect animals, and at the same time not to injure Physiology, and Huxley and Sanderson's approval almost suffices on this head. Pray let me have a line from you soon.

Yours affectionately,

CHARLES DARWIN.

[The Physiological Society, which was founded in 1876, was in some measure the outcome of the anti-vivisection movement, since it was this agitation which impressed on Physiologists the need of a centre for those engaged in this particular branch of science. With respect to the Society, my father wrote to Mr. Romanes (May 29, 1876):—

"I was very much gratified by the wholly unexpected honour of being elected one of the Honorary Members. This mark of sympathy has pleased me to a very high degree."

* Mr. R. B. Litchfield, his son-in-law.

The following letter appeared in the *Times*, April 18th, 1881:]

*C. Darwin to Frithiof Holmgren.**

Down, April 14, 1881.

DEAR SIR,—In answer to your courteous letter of April 7, I have no objection to express my opinion with respect to the right of experimenting on living animals. I use this latter expression as more correct and comprehensive than that of vivisection. You are at liberty to make any use of this letter which you may think fit, but if published I should wish the whole to appear. I have all my life been a strong advocate for humanity to animals, and have done what I could in my writings to enforce this duty. Several years ago, when the agitation against physiologists commenced in England, it was asserted that inhumanity was here practised, and useless suffering caused to animals; and I was led to think that it might be advisable to have an Act of Parliament on the subject. I then took an active part in trying to get a Bill passed, such as would have removed all just cause of complaint, and at the same time have left physiologists free to pursue their researches,—a Bill very different from the Act which has since been passed. It is right to add that the investigation of the matter by a Royal Commission proved that the accusations made against our English physiologists were false. From all that I have heard, however, I fear that in some parts of Europe little regard is paid to the sufferings of animals, and if this be the case, I should be glad to hear of legislation against inhumanity in any such country. On the other hand, I know that physiology cannot possibly progress except by means of experiments on living animals, and I feel the deepest conviction that he who retards the progress of physiology commits a crime against mankind. Any one

* Professor of Physiology at Upsala.

who remembers, as I can, the state of this science half a century ago, must admit that it has made immense progress, and it is now progressing at an ever-increasing rate. What improvements in medical practice may be directly attributed to physiological research is a question which can be properly discussed only by those physiologists and medical practitioners who have studied the history of their subjects; but, as far as I can learn, the benefits are already great. However this may be, no one, unless he is grossly ignorant of what science has done for mankind, can entertain any doubt of the incalculable benefits which will hereafter be derived from physiology, not only by man, but by the lower animals. Look for instance at Pasteur's results in modifying the germs of the most malignant diseases, from which, as it so happens, animals will in the first place receive more relief than man. Let it be remembered how many lives and what a fearful amount of suffering have been saved by the knowledge gained of parasitic worms through the experiments of Virchow and others on living animals. In the future every one will be astonished at the ingratitude shown, at least in England, to these benefactors of mankind. As for myself, permit me to assure you that I honour, and shall always honour, every one who advances the noble science of physiology.

Dear Sir, yours faithfully,

CHARLES DARWIN.

[In the *Times* of the following day appeared a letter headed "Mr. Darwin and Vivisection," signed by Miss Frances Power Cobbe. To this my father replied in the *Times* of April 22, 1881. On the same day he wrote to Mr. Romanes:—

"As I have a fair opportunity, I sent a letter to the *Times* on Vivisection, which is printed to-day. I thought it fair to bear my share of the abuse poured in so atrocious a manner on all physiologists."]

C. Darwin to the Editor of the 'Times.'

SIR,—I do not wish to discuss the views expressed by Miss Cobbe in the letter which appeared in the *Times* of the 19th inst.; but as she asserts that I have "misinformed" my correspondent in Sweden in saying that "the investigation of the matter by a Royal Commission proved that the accusations made against our English physiologists were false," I will merely ask leave to refer to some other sentences from the report of the Commission.

(1.) The sentence—"It is not to be doubted that inhumanity may be found in persons of very high position as physiologists," which Miss Cobbe quotes from page 17 of the report, and which, in her opinion, "can necessarily concern English physiologists alone and not foreigners," is immediately followed by the words "We have seen that it was so in Majendie." Majendie was a French physiologist who became notorious some half century ago for his cruel experiments on living animals.

(2.) The Commissioners, after speaking of the "general sentiment of humanity" prevailing in this country, say (p. 10):—

"This principle is accepted generally by the very highly educated men whose lives are devoted either to scientific investigation and education or to the mitigation or the removal of the sufferings of their fellow-creatures; though differences of degree in regard to its practical application will be easily discernible by those who study the evidence as it has been laid before us."

Again, according to the Commissioners (p. 10):—

"The secretary of the Royal Society for the Prevention of Cruelty to Animals, when asked whether the general tendency of the scientific world in this country is at variance with humanity, says he believes it to be very different, indeed, from that of foreign physiologists; and while giving it as the opinion of the society that experiments are performed which are in their nature beyond any legitimate province of science, and that the pain which they inflict is pain which it is not justifiable to inflict even for the scientific object in view, he readily acknowledges that he does not know a single case of wanton cruelty, and that in general the English physiologists have used anæsthetics where they think they can do so with safety to the experiment."

I am, Sir, your obedient servant,

CHARLES DARWIN.

April 21.

[In the *Times* of Saturday, April 23, 1881, appeared a letter from Miss Cobbe in reply.]

C. Darwin to G. J. Romanes.

Down, April 25, 1881.

MY DEAR ROMANES,—I was very glad to read your last note with much news interesting to me. But I write now to say how I, and indeed all of us in the house, have admired your letter in the *Times*.* It was so simple and direct. I was particularly glad about Burdon Sanderson, of whom I have been for several years a great admirer. I was also especially glad to read the last sentences. I have been bothered with several letters, but none abusive. Under a *selfish* point of view I am very glad of the publication of your letter, as I was at first inclined to think that I had done mischief by stirring up the mud. Now I feel sure that I have done good. Mr. Jesse has written to me very politely, he says his Society has had nothing to do with placards and diagrams against physiology, and I suppose, therefore, that these all originate with Miss Cobbe. . . . Mr. Jesse complains bitterly that the

* April 25, 1881.—Mr. Romanes defended Dr. Sanderson against the accusations made by Miss Cobbe.

Times will "burke" all his letters to this newspaper, nor am I surprised, judging from the laughable tirades advertised in 'Nature.'

Ever yours, very sincerely,
CH. DARWIN.

[The next letter refers to a projected conjoint article on vivisection, to which Mr. Romanes wished my father to contribute:]

C. Darwin to G. J. Romanes.

Down, September 2, 1881.

MY DEAR ROMANES,—Your letter has perplexed me beyond all measure. I fully recognise the duty of every one whose opinion is worth anything, expressing his opinion publicly on vivisection; and this made me send my letter to the *Times*. I have been thinking at intervals all morning what I could say, and it is the simple truth that I have nothing worth saying. You and men like you, whose ideas flow freely, and who can express them easily, cannot understand the state of mental paralysis in which I find myself. What is most wanted is a careful and accurate attempt to show what physiology has already done for man, and even still more strongly what there is every reason to believe it will hereafter do. Now I am absolutely incapable of doing this, or of discussing the other points suggested by you.

If you wish for my name (and I should be glad that it should appear with that of others in the same cause), could you not quote some sentence from my letter in the *Times* which I enclose, but please return it. If you thought fit you might say you quoted it with my approval, and that after still further reflection I still abide most strongly in my expressed conviction.

For Heaven's sake, do think of this. I do not grudge the labour and thought; but I could write nothing worth any one reading.

Allow me to demur to your calling your conjoint article a "symposium" strictly a "drinking party." This seems to me very bad taste, and I do hope every one of you will avoid any semblance of a joke on the subject. I *know* that words, like a joke, on this subject have quite disgusted some persons not at all inimical to physiology. One person lamented to me that Mr. Simon, in his truly admirable Address at the Medical Congress (by far the best thing which I have read), spoke of the fantastic *sensuality* * (or some such term) of the many mistaken, but honest men and women who are half mad on the subject. . . .

[To Dr. Lauder Brunton my father wrote in February 1882:—

"Have you read Mr. [Edmund] Gurney's articles in the 'Fortnightly' † and 'Cornhill'? ‡ They seem to me very clever, though obscurely written, and I agree with almost everything he says, except with some passages which appear to imply that no experiments should be tried unless some immediate good can be predicted, and this is a gigantic mistake contradicted by the whole history of science."]

* 'Transactions of the International Medical Congress,' 1881, vol. iv. p. 413. The expression "lackadaisical" (not fantastic), and "feeble sensuality," are used with regard to the feelings of the anti-vivisectionists.

† "A chapter in the Ethics of Pain," 'Fortnightly Review,' 1881, vol. xxx. p. 778.

‡ "An Epilogue on Vivisection," 'Cornhill Magazine,' 1882, vol. xlv. p. 191.

CHAPTER VI.

MISCELLANEA (*continued*)—A REVIVAL OF GEOLOGICAL WORK—THE BOOK ON EARTHWORMS—LIFE OF ERASMUS DARWIN—MISCELLANEOUS LETTERS.

1876–1882.

[WE have now to consider the work (other than botanical) which occupied the concluding six years of my father's life. A letter to his old friend Rev. L. Blomfield (Jenyns), written in March, 1877, shows what was my father's estimate of his own powers of work at this time :—

"MY DEAR JENYNS (I see I have forgotten your proper names),—Your extremely kind letter has given me warm pleasure. As one gets old, one's thoughts turn back to the past rather than to the future, and I often think of the pleasant, and to me valuable, hours which I spent with you on the borders of the Fens.

"You ask about my future work; I doubt whether I shall be able to do much more that is new, and I always keep before my mind the example of poor old ——, who in his old age had a cacoethes for writing. But I cannot endure doing nothing, so I suppose that I shall go on as long as I can without obviously making a fool of myself. I have a great mass of matter with respect to variation under nature; but so much has been published since the appearance of the 'Origin of Species,' that I very much doubt whether I retain power of mind and strength to reduce the mass into a digested whole. I have sometimes thought that I would try, but dread the attempt. . . ."

His prophecy proved to be a true one with regard to any continuation of any general work in the direction of Evolution, but his estimate of powers which could afterwards prove capable of grappling with the 'Movements of Plants,' and with the work on 'Earthworms,' was certainly a low one.

The year 1876, with which the present chapter begins, brought with it a revival of geological work. He had been astonished, as I hear from Professor Judd, and as appears in his letters, to learn that his books on 'Volcanic Islands,' 1844, and on 'South America,' 1846, were still consulted by geologists, and it was a surprise to him that new editions should be required. Both these works were originally published by Messrs. Smith and Elder, and the new edition of 1876 was also brought out by them. This appeared in one volume with the title 'Geological Observations on the Volcanic Islands, and Parts of South America visited during the Voyage of H.M.S. *Beagle*.' He has explained in the preface his reasons for leaving untouched the text of the original editions: "They relate to parts of the world which have been so rarely visited by men of science, that I am not aware that much could be corrected or added from observations subsequently made. Owing to the great progress which Geology has made within recent times, my views on some few points may be somewhat antiquated; but I have thought it best to leave them as they originally appeared."

It may have been the revival of geological speculation, due to the revision of his early books, that led to his recording the observations of which some account is given in the following letter. Part of it has been published in Professor James Geikie's 'Prehistoric Europe,' chaps. vii. and ix.,* a few verbal alterations having been made at my father's request in the passages quoted. Mr. Geikie lately wrote to me: "The

* My father's suggestion is also noticed in Prof. Geikie's address on the 'Ice Age in Europe and North America,' given at Edinburgh, Nov. 20, 1884.

views suggested in his letter as to the origin of the angular gravels, &c., in the South of England will, I believe, come to be accepted as the truth. This question has a much wider bearing than might at first appear. In point of fact it solves one of the most difficult problems in Quaternary Geology—and has already attracted the attention of German geologists."]

C. Darwin to James Geikie.

Down, November 16, 1876.

My dear Sir,—I hope that you will forgive me for troubling you with a very long letter. But first allow me to tell you with what extreme pleasure and admiration I have just finished reading your 'Great Ice Age.' It seems to me admirably done, and most clear. Interesting as many chapters are in the history of the world, I do not think that any one comes [up] nearly to the glacial period or periods. Though I have steadily read much on the subject, your book makes the whole appear almost new to me.

I am now going to mention a small observation, made by me two or three years ago, near Southampton, but not followed out, as I have no strength for excursions. I need say nothing about the character of the drift there (which includes palæolithic celts), for you have described its essential features in a few words at p. 506. It covers the whole country [in an] even plain-like surface, almost irrespective of the present outline of the land.

The coarse stratification has sometimes been disturbed. I find that you allude "to the larger stones often standing on end;" and this is the point which struck me so much. Not only moderately sized angular stones, but small oval pebbles often stand vertically up, in a manner which I have never seen in ordinary gravel beds. This fact reminded me of what occurs near my home, in the stiff red clay, full of unworn flints over the chalk, which is no doubt the residue left undissolved by rain water. In this clay, flints as long and thin as my arm often stand perpendicularly up; and I have been told by the tank-diggers that it is their "natural position"! I presume that this position may safely be attributed to the differential movement of parts of the red clay as it subsided very slowly from the dissolution of the underlying chalk; so that the flints arrange themselves in the lines of least resistance. The similar but less strongly marked arrangement of the stones in the drift near Southampton makes me suspect that it also must have slowly subsided; and the notion has crossed my mind that during the commencement and height of the glacial period great beds of frozen snow accumulated over the south of England, and that, during the summer, gravel and stones were washed from the higher land over its surface, and in superficial channels. The larger streams may have cut right through the frozen snow, and deposited gravel in lines at the bottom. But on each succeeding autumn, when the running water failed, I imagine that the lines of drainage would have been filled up by blown snow afterwards congealed, and that, owing to great surface accumulations of snow, it would be a mere chance whether the drainage, together with gravel and sand, would follow the same lines during the next summer. Thus, as I apprehend, alternate layers of frozen snow and drift, in sheets and lines, would ultimately have covered the country to a great thickness, with lines of drift probably deposited in various directions at the bottom by the larger streams. As the climate became warmer, the lower beds of frozen snow would have melted with extreme slowness, and the many irregular beds of interstratified drift would have sunk down with equal slowness; and during this movement the elongated pebbles would have arranged themselves more or less vertically. The drift would also have been deposited almost irrespective of the outline of the underlying land. When I viewed the country I could not persuade myself that any flood, however great, could have deposited such coarse gravel over the almost level platforms between the valleys. My view differs from that of Holst, p. 415 ['Great Ice Age'], of which I had never heard, as his relates to channels cut through glaciers, and mine to beds of drift interstratified with frozen snow where no glaciers existed. The upshot of this long letter is to ask you to keep my notion in your head, and look out for upright pebbles in any lowland country which you may examine, where glaciers have not existed. Or if you think the notion deserves any further thought, but not otherwise, to tell any one of it, for instance Mr. Skertchly, who is examining such districts. Pray forgive me for writing so long a letter, and again thanking you for the great pleasure derived from your book,

I remain yours very faithfully,

CH. DARWIN.

P.S. . . . I am glad that you have read Blytt;* his paper seemed to me a most important contribution to Botanical Geography. How curious that the same conclusions should have been arrived at by Mr. Skertchly, who seems to be a first-rate observer; and this implies, as I always think, a sound theoriser.

I have told my publisher to send you in two or three days a copy (second edition) of my geological work during the voyage of the *Beagle*. The sole point which would perhaps interest you is about the steppe-like plains of Patagonia.

For many years past I have had fearful misgivings that it must have been the level of the sea, and not that of the land which has changed.

I read a few months ago your [brother's] very interesting life of Murchison.† Though I have always thought that he ranked next to W. Smith in the classification of formations,

* Axel Blytt.—'Essay on the Immigration of the Norwegian Flora during alternate rainy and dry Seasons.' Christiania, 1876.

† By Mr. Archibald Geikie.

and though I knew how kind-hearted [he was], yet the book has raised him greatly in my respect, notwithstanding his foibles and want of broad philosophical views.

[The only other geological work of his later years was embodied in his book on earthworms (1881), which may therefore be conveniently considered in this place. This subject was one which had interested him many years before this date, and in 1838 a paper on the formation of mould was published in the Proceedings of the Geological Society (see vol. i. p. 284).

Here he showed that "fragments of burnt marl, cinders, &c., which had been thickly strewed over the surface of several meadows were found after a few years lying at a depth of some inches beneath the turf, but still forming a layer." For the explanation of this fact, which forms the central idea of the geological part of the book, he was indebted to his uncle Josiah Wedgwood, who suggested that worms, by bringing earth to the surface in their castings, must undermine any objects lying on the surface and cause an apparent sinking.

In the book of 1881 he extended his observations on this burying action, and devised a number of different ways of checking his estimates as to the amount of work done.* He also added a mass of observations on the habits, natural history and intelligence of worms, a part of the work which added greatly to its popularity.

In 1877 Sir Thomas Farrer had discovered close to his garden the remains of a building of Roman-British times, and thus gave my father the opportunity of seeing for himself

* He received much valuable help from Dr. King, of the Botanical Gardens, Calcutta. The following passage is from a letter to Dr. King, dated January 18, 1873:—

"I really do not know how to thank you enough for the immense trouble which you have taken. You have attended *exactly* and *fully* to the points about which I was most anxious. If I had been each evening by your side, I could not have suggested anything else."

the effects produced by earthworms on the old concrete-floors, walls, &c. On his return he wrote to Sir Thomas Farrer :—

"I cannot remember a more delightful week than the last. I know very well that E. will not believe me, but the worms were by no means the sole charm."

In the autumn of 1880, when the 'Power of Movements in Plants' was nearly finished, he began once more on the subject. He wrote to Professor Carus (September 21) :—

"In the intervals of correcting the press, I am writing a very little book, and have done nearly half of it. Its title will be (as at present designed), 'The Formation of Vegetable Mould through the Action of Worms.'* As far as I can judge it will be a curious little book."

The manuscript was sent to the printers in April, 1881, and when the proof-sheets were coming in he wrote to Professor Carus : "The subject has been to me a hobby-horse, and I have perhaps treated it in foolish detail."

It was published on October 10, and 2000 copies were sold at once. He wrote to Sir J. D. Hooker, "I am glad that you approve of the 'Worms.' When in old days I was to tell you whatever I was doing, if you were at all interested, I always felt as most men do when their work is finally published."

To Mr. Mellard Reade he wrote (November 8) : "It has been a complete surprise to me how many persons have cared for the subject." And to Mr. Dyer (in November) : "My book has been received with almost laughable enthusiasm, and 3500 copies have been sold ! ! !" Again, to his friend Mr. Anthony Rich, he wrote on February 4, 1882, "I have been plagued with an endless stream of letters on the subject ; most of them very foolish and enthusiastic ; but some containing good facts which I have used in correcting yesterday the 'Sixth Thousand.'" The popularity of the

* The full title is 'The Formation of Vegetable Mould through the Action of Worms, with Observations on their Habits,' 1881.

book may be roughly estimated by the fact that, in the three years following its publication, 8500 copies were sold—a sale relatively greater than that of the 'Origin of Species.'

It is not difficult to account for its success with the non-scientific public. Conclusions so wide and so novel, and so easily understood, drawn from the study of creatures so familiar, and treated with unabated vigour and freshness, may well have attracted many readers. A reviewer remarks: "In the eyes of most men. . . the earthworm is a mere blind, dumb, senseless, and unpleasantly slimy annelid. Mr. Darwin undertakes to rehabilitate his character, and the earthworm steps forth at once as an intelligent and beneficent personage, a worker of vast geological changes, a planer down of mountain sides . . . a friend of man . . . and an ally of the Society for the preservation of ancient monuments." The *St. James's Gazette*, of October 17th, 1881, pointed out that the teaching of the cumulative importance of the infinitely little is the point of contact between this book and the author's previous work.

One more book remains to be noticed, the 'Life of Erasmus Darwin.'

In February 1879 an essay by Dr. Ernst Krause, on the scientific work of Erasmus Darwin, appeared in the evolutionary journal, 'Kosmos,' The number of 'Kosmos' in question was a "Gratulationsheft," * or special congratulatory issue in honour of my father's birthday, so that Dr. Krause's essay, glorifying the older evolutionist, was quite in its place. He wrote to Dr. Krause, thanking him cordially for the honour paid to Erasmus, and asking his permission to publish † an English translation of the Essay.

* The same number contains a good biographical sketch of my father, of which the material was to a large extent supplied by him to the writer, Prof. Preyer of Jena. The article contains an excellent list of my father's publications.

† The wish to do so was shared by his brother, Erasmus Darwin the younger, who continued to be associated with the project.

His chief reason for writing a notice of his grandfather's life was "to contradict flatly some calumnies by Miss Seward." This appears from a letter of March 27, 1879, to his cousin Reginald Darwin, in which he asks for any documents and letters which might throw light on the character of Erasmus. This led to Mr. Reginald Darwin placing in my father's hands a quantity of valuable material, including a curious folio common-place book, of which he wrote: "I have been deeply interested by the great book, ... reading and looking at it is like having communion with the dead .... [it] has taught me a good deal about the occupations and tastes of our grandfather." A subsequent letter (April 8) to the same correspondent describes the source of a further supply of material:—

"Since my last letter I have made a strange discovery; for an old box from my father marked 'Old Deeds,' and which consequently I had never opened, I found full of letters—hundreds from Dr. Erasmus—and others from old members of the family: some few very curious. Also a drawing of Elston before it was altered, about 1750, of which I think I will give a copy."

Dr. Krause's contribution formed the second part of the 'Life of Erasmus Darwin,' my father supplying a "preliminary notice." This expression on the title-page is somewhat misleading; my father's contribution is more than half the book, and should have been described as a biography. Work of this kind was new to him, and he wrote doubtfully to Mr. Thiselton Dyer, June 18th: "God only knows what I shall make of his life, it is such a new kind of work to me." The strong interest he felt about his forebears helped to give zest to the work, which became a decided enjoyment to him. With the general public the book was not markedly successful, but many of his friends recognised its merits. Sir J. D. Hooker was one of these, and to him my father wrote, "Your praise of the Life of Dr. D. has pleased me exceedingly, for I despised my work, and thought myself a perfect fool to have undertaken such a job."

To Mr. Galton, too, he wrote, November 14:—

"I am *extremely* glad that you approve of the little 'Life' of our grandfather, for I have been repenting that I ever undertook it, as the work was quite beyond my tether."

The publication of the 'Life of Erasmus Darwin' led to an attack by Mr. Samuel Butler, which amounted to a charge of falsehood against my father. After consulting his friends, he came to the determination to leave the charge unanswered. as being unworthy of his notice.* Those who wish to know more of the matter, may gather the facts of the case from Ernst Krause's 'Charles Darwin,' and they will find Mr. Butler's statement of his grievance in the *Athenæum*, January 31, 1880, and in the *St. James's Gazette*, December 8, 1880. The affair gave my father much pain, but the warm sympathy of those whose opinion he respected soon helped him to let it pass into a well-merited oblivion.

The following letter refers to M. J. H. Fabre's 'Souvenirs Entomologiques.' It may find a place here, as it contains a defence of Erasmus Darwin on a small point. The postscript is interesting, as an example of one of my father's bold ideas both as to experiment and theory :]

C. Darwin to J. H. Fabre.

Down, January 31, 1880.

MY DEAR SIR,—I hope that you will permit me to have the satisfaction of thanking you cordially for the lively pleasure which I have derived from reading your book. Never have the wonderful habits of insects been more vividly described, and it is almost as good to read about them as to

* He had, in a letter to Mr. Butler, expressed his regret at the oversight which caused so much offence.

see them. I feel sure that you would not be unjust to even an insect, much less to a man. Now, you have been misled by some translator, for my grandfather, Erasmus Darwin, states ('Zoonomia,' vol. i. p. 183, 1794) that it was a wasp (guêpe) which he saw cutting off the wings of a large fly. I have no doubt that you are right in saying that the wings are generally cut off instinctively; but in the case described by my grandfather, the wasp, after cutting off the two ends of the body, rose in the air, and was turned round by the wind; he then alighted and cut off the wings. I must believe, with Pierre Huber, that insects have "une petite dose de raison." In the next edition of your book, I hope that you will alter *part* of what you say about my grandfather.

I am sorry that you are so strongly opposed to the Descent theory; I have found the searching for the history of each structure or instinct an excellent aid to observation; and wonderful observer as you are, it would suggest new points to you. If I were to write on the evolution of instincts, I could make good use of some of the facts which you give. Permit me to add, that when I read the last sentence in your book, I sympathised deeply with you.*

With the most sincere respect,

I remain, dear Sir, yours faithfully,

CHARLES DARWIN.

P.S.—Allow me to make a suggestion in relation to your wonderful account of insects finding their way home. I formerly wished to try it with pigeons: namely, to carry the insects in their paper "cornets," about a hundred paces in the opposite direction to that which you ultimately intended to carry them; but before turning round to return, to put the insect in a circular box, with an axle which could be made to

* The book is intended as a memorial of the early death of M. Fabre's son, who had been his father's assistant in his observations on insect life.

revolve very rapidly, first in one direction, and then in another, so as to destroy for a time all sense of direction in the insects. I have sometimes *imagined* that animals may feel in which direction they were at the first start carried.* If this plan failed, I had intended placing the pigeons within an induction coil, so as to disturb any magnetic or dia-magnetic sensibility, which it seems just possible that they may possess. C. D.

[During the latter years of my father's life there was a growing tendency in the public to do him honour. In 1877 he received the honorary degree of LL.D. from the University of Cambridge. The degree was conferred on November 17, and with the customary Latin speech from the Public Orator, concluding with the words: "Tu vero, qui leges naturæ tam docte illustraveris, legum doctor nobis esto."

The honorary degree led to a movement being set on foot in the University to obtain some permanent memorial of my father. A sum of about £400 was subscribed, and after the rejection of the idea that a bust would be the best memorial, a picture was determined on. In June 1879 he sat to Mr. W. Richmond for the portrait in the possession of the University, now placed in the Library of the Philosophical Society at Cambridge. He is represented seated in a Doctor's gown, the head turned towards the spectator: the picture has many admirers, but, according to my own view, neither the attitude nor the expression are characteristic of my father.

A similar wish on the part of the Linnean Society—with which my father was so closely associated—led to his sitting

* This idea was a favourite one with him, and he has described in 'Nature' (vol. vii. 1873, p. 360) the behaviour of his cob Tommy, in whom he fancied he detected a sense of direction. The horse had been taken by rail from Kent to the Isle of Wight; when there he exhibited a marked desire to go eastward, even when his stable lay in the opposite direction. In the same volume of 'Nature,' p. 417, is a letter on the 'Origin of Certain Instincts,' which contains a short discussion on the sense of direction.

in August, 1881, to Mr. John Collier, for the portrait now in the possession of the Society. Of the artist, he wrote, 'Collier was the most considerate, kind and pleasant painter a sitter could desire." The portrait represents him standing facing the observer in the loose cloak so familiar to those who knew him, and with his slouch hat in his hand. Many of those who knew his face most intimately, think that Mr. Collier's picture is the best of the portraits, and in this judgment the sitter himself was inclined to agree. According to my feeling it is not so simple or strong a representation of him as that given by Mr. Ouless. There is a certain expression in Mr. Collier's portrait which I am inclined to consider an exaggeration of the almost painful expression which Professor Cohn has described in my father's face, and which he had previously noticed in Humboldt. Professor Cohn's remarks occur in a pleasantly written account of a visit to Down* in 1876, published in the *Breslauer Zeitung*, April 23, 1882.

Besides the Cambridge degree, he received about the same time honours of an academic kind from some foreign societies.

On August 5, 1878, he was elected a Corresponding Member of the French Institute† in the Botanical Section,‡ and wrote to Dr. Asa Gray:—

"I see that we are both elected Corresponding Members

* In this connection may be mentioned a visit (1881) from another distinguished German, Hans Richter. The occurrence is otherwise worthy of mention, inasmuch as it led to the publication, after my father's death, of Herr Richter's recollections of the visit. The sketch is simply and sympathetically written, and the author has succeeded in giving a true picture of my father as he lived at Down. It appeared in the *Neue Tagblatt* of Vienna, and was republished by Dr. O. Zacharias in his 'Charles R. Darwin,' Berlin, 1882.

† "Lyell always spoke of it as a great scandal that Darwin was so long kept out of the French Institute. As he said, even if the development hypothesis were objected to, Darwin's original works on Coral Reefs, the Cirripedia, and other subjects, constituted a more than sufficient claim."—From Professor Judd's notes.

‡ The statement has been more than once published that he was elected to the Zoological Section, but this was not the case.

of the Institute. It is rather a good joke that I should be elected in the Botanical Section, as the extent of my knowledge is little more than that a daisy is a Compositous plant and a pea a Leguminous one."

In the early part of the same year he was elected a Corresponding Member of the Berlin Academy of Sciences, and he wrote (March 12) to Professor Du Bois Reymond, who had proposed him for election:—

"I thank you sincerely for your most kind letter, in which you announce the great honour conferred on me. The knowledge of the names of the illustrious men, who seconded the proposal is even a greater pleasure to me than the honour itself."

The seconders were Helmholtz, Peters, Ewald, Pringsheim and Virchow.

In 1879 he received the Baly Medal of the Royal College of Physicians.*

He received twenty-six votes out of a possible 39, five blank papers were sent in, and eight votes were recorded for the other candidates.

In 1872 an attempt had been made to elect him to the Section of Zoology, when, however, he only received 15 out of 48 votes, and Lovén was chosen for the vacant place. It appears ('Nature,' August 1, 1872), that an eminent member of the Academy, wrote to *Les Mondes* to the following effect:—

"What has closed the doors of the Academy to Mr. Darwin is that the science of those of his books which have made his chief title to fame—the 'Origin of Species,' and still more the 'Descent of Man,' is not science, but a mass of assertions and absolutely gratuitous hypotheses, often evidently fallacious. This kind of publication and these theories are a bad example, which a body that respects itself cannot encourage."

* The visit to London, necessitated by the presentation of the Baly Medal, was combined with a visit to Miss Forster's house at Abinger, in Surrey, and this was the occasion of the following characteristic letter:—"I must write a few words to thank you cordially for lending us your house. It was a most kind thought, and has pleased me greatly; but I know well that I do not deserve such kindness from any one. On the other hand, no one can be too kind to my dear wife, who is worth her weight in gold many times over, and she was anxious that I should get some complete rest, and here I cannot rest. Your house will be a delightful haven, and again I thank you truly."

Again in 1879 he received from the Royal Academy of Turin the *Bressa* Prize for the years 1875–78, amounting to the sum of 12,000 francs. In the following year he received on his birthday, as on previous occasions, a kind letter of congratulation from Dr. Dohrn of Naples. In writing (February 15th) to thank him and the other naturalists at the Zoological Station, my father added:—

"Perhaps you saw in the papers that the Turin Society honoured me to an extraordinary degree by awarding me the *Bressa* Prize. Now it occurred to me that if your station wanted some piece of apparatus, of about the value of £100, I should very much like to be allowed to pay for it. Will you be so kind as to keep this in mind, and if any want should occur to you, I would send you a cheque at any time."

I find from my father's accounts that £100 was presented to the Naples Station.

He received also several tokens of respect and sympathy of a more private character from various sources. With regard to such incidents, and to the estimation of the public generally, his attitude may be illustrated by a passage from a letter to Mr. Romanes:*—

"You have indeed passed a most magnificent eulogium upon me, and I wonder that you were not afraid of hearing 'oh! oh!' or some other sign of disapprobation. Many persons think that what I have done in science has been much overrated, and I very often think so myself; but my comfort is that I have never consciously done anything to gain applause. Enough and too much about my dear self."

Among such expressions of regard he valued very highly the two photographic albums received from Germany and Holland on his birthday, 1877. Herr Emil Rade of Münster, originated the idea of the German birthday gift, and under-

* The lecture referred to was given at the Dublin meeting of the British Association.

took the necessary arrangements. To him my father wrote (February 16, 1877):—

"I hope that you will inform the one hundred and fifty-four men of science, including some of the most highly honoured names in the world, how grateful I am for their kindness and generous sympathy in having sent me their photographs on my birthday."

To Professor Haeckel he wrote (February 16, 1877):—

"The album has just arrived quite safe. It is most superb.* It is by far the greatest honour which I have ever received, and my satisfaction has been greatly enhanced, by your most kind letter of February 9. . . . I thank you all from my heart. I have written by this post to Herr Rade, and I hope he will somehow manage to thank all my generous friends."

To Professor A. van Bemmelen he wrote, on receiving a similar present from a number of distinguished men and lovers of Natural History in the Netherlands:—

"SIR,—I received yesterday the magnificent present of the album, together with your letter. I hope that you will endeavour to find some means to express to the two hundred and seventeen distinguished observers and lovers of natural science, who have sent me their photographs, my gratitude for their extreme kindness. I feel deeply gratified by this gift, and I do not think that any testimonial more honourable to me could have been imagined. I am well aware that my books could never have been written, and would not have made any impression on the public mind, had not an immense amount of material been collected by a long series of admirable observers; and it is to them that honour is chiefly due. I suppose that every worker at science occasionally feels depressed, and doubts whether what he has published has been worth the labour which it has cost him, but for the few

* The album is magnificently bound and decorated with a beautifully illuminated titlepage, the work of an artist, Herr A. Fitger of Bremen, who also contributed the dedicatory poem.

remaining years of my life, whenever I want cheering, I will look at the portraits of my distinguished co-workers in the field of science, and remember their generous sympathy. When I die, the album will be a most precious bequest to my children. I must further express my obligation for the very interesting history contained in your letter of the progress of opinion in the Netherlands, with respect to Evolution, the whole of which is quite new to me. I must again thank all my kind friends, from my heart, for their ever-memorable testimonial, and I remain, Sir,

Your obliged and grateful servant,

CHARLES R. DARWIN."

[In the June of the following year (1878) he was gratified by learning that the Emperor of Brazil had expressed a wish to meet him. Owing to absence from home my father was unable to comply with this wish; he wrote to Sir J. D. Hooker:—

"The Emperor has done so much for science, that every scientific man is bound to show him the utmost respect, and I hope that you will express in the strongest language, and which you can do with entire truth, how greatly I feel honoured by his wish to see me; and how much I regret my absence from home."

Finally it should be mentioned that in 1880 he received an address personally presented by members of the Council of the Birmingham Philosophical Society, as well as a memorial from the Yorkshire Naturalist Union presented by some of the members, headed by Dr. Sorby. He also received in the same year a visit from some of the members of the Lewisham and Blackheath Scientific Association,—a visit which was, I think, enjoyed by both guests and host.]

MISCELLANEOUS LETTERS—1876–1882.

[The chief incident of a personal kind (not already dealt with) in the years which we are now considering was the death of his brother Erasmus, who died at his house in Queen Anne Street, on August 26th, 1881. My father wrote to Sir J. D. Hooker (Aug. 30):—

"The death of Erasmus is a very heavy loss to all of us, for he had a most affectionate disposition. He always appeared to me the most pleasant and clearest headed man, whom I have ever known. London will seem a strange place to me without his presence; I am deeply glad that he died without any great suffering, after a very short illness from mere weakness and not from any definite disease.*

"I cannot quite agree with you about the death of the old and young. Death in the latter case, when there is a bright future ahead, causes grief never to be wholly obliterated."

An incident of a happy character may also be selected for especial notice, since it was one which strongly moved my father's sympathy. A letter (Dec. 17, 1879) to Sir Joseph Hooker shows that the possibility of a Government Pension being conferred on Mr. Wallace first occurred to my father at this time. The idea was taken up by others, and my father's letters show that he felt the most lively interest in the success of the plan. He wrote, for instance, to Mrs. Fisher, "I hardly ever wished for anything more than I do for the success of our plan." He was deeply pleased when this thoroughly deserved honour was bestowed on his friend, and wrote to the same correspondent (January 7, 1881), on receiving a letter from Mr. Gladstone announcing the fact: "How extraordinarily kind of Mr. Gladstone to find time to write under

* "He was not, I think, a happy man, and for many years did not value life, though never complaining."—From a letter to Sir Thomas Farrer.

the present circumstances.* Good heavens! how pleased I am!"

The letters which follow are of a miscellaneous character and refer principally to the books he read, and to his minor writings.]

C. Darwin to Miss Buckley (Mrs. Fisher).

Down, February 11 [1876].

MY DEAR MISS BUCKLEY,—You must let me have the pleasure of saying that I have just finished reading with very great interest your new book.† The idea seems to me a capital one, and as far as I can judge very well carried out. There is much fascination in taking a bird's eye view of all the grand leading steps in the progress of science. At first I regretted that you had not kept each science more separate; but I dare say you found it impossible. I have hardly any criticisms, except that I think you ought to have introduced Murchison as a great classifier of formations, second only to W. Smith. You have done full justice, and not more than justice, to our dear old master, Lyell. Perhaps a little more ought to have been said about botany, and if you should ever add this, you would find Sachs' 'History,' lately published, very good for your purpose.

You have crowned Wallace and myself with much honour and glory. I heartily congratulate you on having produced so novel and interesting a work, and remain,

My dear Miss Buckley, yours very faithfully,

CH. DARWIN.

* Mr. Gladstone was then in office, and the letter must have been written when he was overwhelmed with business connected with the opening of Parliament (Jan. 6).

† 'A Short History of Natural Science.'

C. Darwin to A. R. Wallace.

[Hopedene] *, June 5, 1876.

My dear Wallace,—I must have the pleasure of expressing to you my unbounded admiration of your book,† tho' I have read only to page 184—my object having been to do as little as possible while resting. I feel sure that you have laid a broad and safe foundation for all future work on Distribution. How interesting it will be to see hereafter plants treated in strict relation to your views; and then all insects, pulmonate molluscs and fresh-water fishes, in greater detail than I suppose you have given to these lower animals. The point which has interested me most, but I do not say the most valuable point, is your protest against sinking imaginary continents in a quite reckless manner, as was stated by Forbes, followed, alas, by Hooker, and caricatured by Wollaston and [Andrew] Murray! By the way, the main impression that the latter author has left on my mind is his utter want of all scientific judgment. I have lifted up my voice against the above view with no avail, but I have no doubt that you will succeed, owing to your new arguments and the coloured chart. Of a special value, as it seems to me, is the conclusion that we must determine the areas, chiefly by the nature of the mammals. When I worked many years ago on this subject, I doubted much whether the now called Palæarctic and Nearctic regions ought to be separated; and I determined if I made another region that it should be Madagascar. I have, therefore, been able to appreciate your evidence on these points. What progress Palæontology has made during the last 20 years; but if it advances at the same rate in the future, our views on the migration and birth-place of the various groups will, I fear, be greatly altered. I cannot feel quite easy about the Glacial period, and the extinction of large

* Mr. Hensleigh Wedgwood's house in Surrey.

† 'Geographical Distribution,' 1876.

mammals, but I must hope that you are right. I think you will have to modify your belief about the difficulty of dispersal of land molluscs; I was interrupted when beginning to experimentize on the just hatched young adhering to the feet of ground-roosting birds. I differ on one other point, viz. in the belief that there must have existed a Tertiary Antarctic continent, from which various forms radiated to the southern extremities of our present continents. But I could go on scribbling for ever. You have written, as I believe, a grand and memorable work which will last for years as the foundation for all future treatises on Geographical Distribution.

My dear Wallace, yours very sincerely,

CHARLES DARWIN.

P.S.—You have paid me the highest conceivable compliment, by what you say of your work in relation to my chapters on distribution in the 'Origin,' and I heartily thank you for it.

[The following letters illustrate my father's power of taking a vivid interest in work bearing on Evolution, but unconnected with his own special researches at the time. The books referred to in the first letter are Professor Weismann's 'Studien zur Descendenzlehre,'* being part of the series of essays by which the author has done such admirable service to the cause of Evolution:]

C. Darwin to Aug. Weismann.

. . . I read German so slowly, and have had lately to read several other papers, so that I have as yet finished only half of your first essay and two-thirds of your second. They have excited my interest and admiration in the highest degree, and whichever I think of last, seems to me the most

* My father contributed a prefatory note to Mr. Meldola's translation of Prof. Weismann's 'Studien,' 1880–81.

valuable. I never expected to see the coloured marks on caterpillars so well explained; and the case of the ocelli delights me especially. . . .

. . . There is one other subject which has always seemed to me more difficult to explain than even the colours of caterpillars, and that is the colour of birds' eggs, and I wish you would take this up.

C. Darwin to Melchior Neumayr, Vienna.*

Down, Beckenham, Kent, March 9, 1877.

DEAR SIR,—From having been obliged to read other books, I finished only yesterday your essay on 'Die Congerien,' &c.†

I hope that you will allow me to express my gratitude for the pleasure and instruction which I have derived from reading it. It seems to me to be an admirable work; and is by far the best case which I have ever met with, showing the direct influence of the conditions of life on the organization.

Mr. Hyatt, who has been studying the Hilgendorf case, writes to me with respect to the conclusions at which he has arrived, and these are nearly the same as yours. He insists that closely similar forms may be derived from distinct lines of descent; and this is what I formerly called analogical variation. There can now be no doubt that species may become greatly modified through the direct action of the environment. I have some excuse for not having formerly insisted more strongly on this head in my 'Origin of Species,' as most of the best facts have been observed since its publication.

With my renewed thanks for your most interesting essay, and with the highest respect, I remain, dear Sir,

Yours very faithfully,

CHARLES DARWIN.

* Professor of Palæontology at Vienna.

† 'Die Congerien und Paludinenschichten Slavoniens,' 4to, 1875.

C. Darwin to E. S. Morse.

Down, April 23, 1877.

My dear Sir,—You must allow me just to tell you how very much I have been interested with the excellent Address * which you have been so kind as to send me, and which I had much wished to read. I believe that I had read all, or very nearly all, the papers by your countrymen to which you refer, but I have been fairly astonished at their number and importance when seeing them thus put together. I quite agree about the high value of Mr. Allen's works,† as showing how much change may be expected apparently through the direct action of the conditions of life. As for the fossil remains in the West, no words will express how wonderful they are. There is one point which I regret that you did not make clear in your Address, namely what is the meaning and importance of Professors Cope and Hyatt's views on acceleration and retardation. I have endeavoured, and given up in despair, the attempt to grasp their meaning.

Permit me to thank you cordially for the kind feeling shown towards me through your Address, and I remain, my dear Sir,

Yours faithfully,

Ch. Darwin.

[The next letter refers to his 'Biographical Sketch of an Infant,' written from notes made 37 years previously, and published in 'Mind,' July, 1877. The article attracted a good deal of attention, and was translated at the time in 'Kosmos,' and the 'Revue Scientifique,' and has been recently published in Dr. Krause's 'Gesammelte kleinere Schriften von Charles Darwin,' 1887 :]

* "What American Zoologists have done for Evolution," an Address to the American Association for the Advancement of Science, August, 1876. Vol. xxv. of the Proceedings of the Association.

† Mr. J. A. Allen shows the existence of geographical races of birds and mammals. Proc. Boston Soc. Nat. Hist. vol. xv.

*C. Darwin to G. Croom Robertson.**

Down, April 27, 1877.

DEAR SIR,—I hope that you will be so good as to take the trouble to read the enclosed MS., and if you think it fit for publication in your admirable journal of 'Mind,' I shall be gratified. If you do not think it fit, as is very likely, will you please to return it to me. I hope that you will read it in an extra critical spirit, as I cannot judge whether it is worth publishing from having been so much interested in watching the dawn of the several faculties in my own infant. I may add that I should never have thought of sending you the MS., had not M. Taine's article appeared in your Journal.† If my MS. is printed, I think that I had better see a proof. I remain, dear Sir,

Yours faithfully,

CH. DARWIN.

[The two following extracts show the lively interest he preserved in diverse fields of inquiry. Professor Cohn, of Breslau, had mentioned, in a letter, Koch's researches on Splenic Fever; my father replied, January 3 :—

"I well remember saying to myself, between twenty and thirty years ago, that if ever the origin of any infectious disease could be proved, it would be the greatest triumph to science; and now I rejoice to have seen the triumph."

In the spring he received a copy of Dr. E. von Mojsisovics' 'Dolomit Riffe;' his letter to the author (June 1, 1878) is interesting, as bearing on the influence of his own work on the methods of geology.

"I have at last found time to read the first chapter of your 'Dolomit Riffe,' and have been *exceedingly* interested by it. What a wonderful change in the future of geological chronology you indicate, by assuming the descent theory to be

* The editor of 'Mind.'

† 1877, p. 252. The original appeared in the 'Revue Philosophique," 1876.

established, and then taking the graduated changes of the same group of organisms as the true standard! I never hoped to live to see such a step even proposed by any one."

Another geological research which roused my father's admiration was Mr. D. Mackintosh's work on erratic blocks. Apart from its intrinsic merit the work keenly excited his sympathy from the conditions under which it was executed, Mr. Mackintosh being compelled to give nearly his whole time to tuition. The following passage is from a letter to Mr. Mackintosh of October 9, 1879, and refers to his paper in the Journal of the Geological Society, 1878:—

"I hope that you will allow me to have the pleasure of thanking you for the very great pleasure which I have derived from just reading your paper on erratic blocks. The map is wonderful, and what labour each of those lines shows! I have thought for some years that the agency of floating ice, which nearly half a century ago was overrated, has of late been underrated. You are the sole man who has ever noticed the distinction suggested by me * between flat or planed scored rocks, and mammillated scored rocks."]

C. Darwin to C. Ridley.

Down, November 28, 1878.

DEAR SIR,—I just skimmed through Dr. Pusey's sermon, as published in the *Guardian*, but it did [not] seem to me worthy of any attention. As I have never answered criticisms excepting those made by scientific men, I am not willing that this letter should be published; but I have no objection to your saying that you sent me the three questions, and that I answered that Dr. Pusey was mistaken in imagining that I wrote the 'Origin' with any relation whatever to Theology. I should have thought that this would have been evident to

* In his paper on the 'Ancient Glaciers of Carnarvonshire,' Phil. Mag. xxi. 1842. See p. 187.

any one who had taken the trouble to read the book, more especially as in the opening lines of the introduction I specify how the subject arose in my mind. This answer disposes of your two other questions; but I may add that, many years ago, when I was collecting facts for the 'Origin,' my belief in what is called a personal God was as firm as that of Dr. Pusey himself, and as to the eternity of matter I have never troubled myself about such insoluble questions. Dr. Pusey's attack will be as powerless to retard by a day the belief in Evolution, as were the virulent attacks made by divines fifty years ago against Geology, and the still older ones of the Catholic Church against Galileo, for the public is wise enough always to follow Scientific men when they agree on any subject; and now there is almost complete unanimity amongst Biologists about Evolution, though there is still considerable difference as to the means, such as how far natural selection has acted, and how far external conditions, or whether there exists some mysterious innate tendency to perfectibility. I remain, dear Sir,

Yours faithfully,

CH. DARWIN.

[Theologians were not the only adversaries of freedom in science. On Sept. 22, 1877, Prof. Virchow delivered an address at the Munich meeting of German Naturalists and Physicians, which had the effect of connecting Socialism with the Descent theory. This point of view was taken up by anti-evolutionists to such an extent that, according to Haeckel, the *Kreuz Zeitung* threw "all the blame" of the "treasonable attempts of the democrats Hödel and Nobiling . . . directly on the theory of Descent." Prof. Haeckel replied with vigour and ability in his 'Freedom in Science and Teaching' (Eng. Transl. 1879), an essay which must have the sympathy of all lovers of freedom.

The following passage from a letter (December 26, 1879) to

Dr. Scherzer, the author of the 'Voyage of the *Novara*,' gives a hint of my father's views on this once burning question:—

"What a foolish idea seems to prevail in Germany on the connection between Socialism and Evolution through Natural Selection."]

*C. Darwin to H. N. Moseley.**

Down, January 20, 1879.

DEAR MOSELEY,—I have just received your book, and I declare that never in my life have I seen a dedication which I admired so much.† Of course I am not a fair judge, but I hope that I speak dispassionately, though you have touched me in my very tenderest point, by saying that my old Journal mainly gave you the wish to travel as a Naturalist. I shall begin to read your book this very evening, and am sure that I shall enjoy it much.

Yours very sincerely,
CH. DARWIN.

C. Darwin to H. N. Moseley.

Down, February 4, 1879.

DEAR MOSELEY,—I have at last read every word of your book, and it has excited in me greater interest than any other scientific book which I have read for a long time. You will perhaps be surprised how slow I have been, but my head prevents me reading except at intervals. If I were asked which parts have interested me most, I should be somewhat

* Professor of Zoology at Oxford. The book alluded to is Prof. Moseley's 'Notes by a Naturalist on the *Challenger*.'

† "To Charles Darwin, Esquire, LL.D., F.R.S., &c., from the study of whose 'Journal of Researches' I mainly derived my desire to travel round the world; to the development of whose theory I owe the principal pleasures and interests of my life, and who has personally given me much kindly encouragement in the prosecution of my studies, this book is, by permission, gratefully dedicated."

puzzled to answer. I fancy that the general reader would prefer your account of Japan. For myself I hesitate between your discussions and description of the Southern ice, which seems to me admirable, and the last chapter which contained many facts and views new to me, though I had read your papers on the stony Hydroid Corals, yet your *résumé* made me realise better than I had done before, what a most curious case it is.

You have also collected a surprising number of valuable facts bearing on the disposal of plants, far more than in any other book known to me. In fact your volume is a mass of interesting facts and discussions, with hardly a superfluous word ; and I heartily congratulate you on its publication.

Your dedication makes me prouder than ever.

Believe me, yours sincerely,

CH. DARWIN.

[In November, 1879, he answered for Mr. Galton a series of questions for his 'Inquiries into Human Faculty,' 1883. He wrote to Mr. Galton :—

"I have answered the questions as well as I could, but they are miserably answered, for I have never tried looking into my own mind. Unless others answer very much better than I can do, you will get no good from your queries. Do you not think you ought to have the age of the answerer? I think so, because I can call up faces of many schoolboys, not seen for sixty years, with *much distinctness*, but nowadays I may talk with a man for an hour, and see him several times consecutively, and, after a month, I am utterly unable to recollect what he is at all like. The picture is quite washed out."

The greater number of the answers are given in the annexed table :]

Questions on the Faculty of Visualising.

	QUESTIONS.	REPLIES.
1	*Illumination?*	Moderate, but my solitary breakfast was early, and the morning dark.
2	*Definition?*	Some objects quite defined, a slice of cold beef, some grapes and a pear, the state of my plate when I had finished, and a few other objects, are as distinct as if I had photos before me.
3	*Completeness?*	Very moderately so.
4	*Colouring?*	The objects above-named, perfectly coloured.
5	*Extent of Field of View.*	Rather small.
	DIFFERENT KINDS OF IMAGERY.	
6	*Printed pages?*	I cannot remember a single sentence, but I remember the place of the sentence and the kind of type.
7	*Furniture?*	I have never attended to it.
8	*Persons?*	I remember the faces of persons formerly well-known vividly, and can make them do anything I like.
9	*Scenery?*	Remembrance vivid and distinct, and gives me pleasure.
10	*Geography?*	No.
11	*Military movements?*	No.
12	*Mechanism?*	Never tried.
13	*Geometry?*	I do not think I have any power of the kind.
14	*Numerals?*	When I think of any number, printed figures arise before my mind. I can't remember for an hour four consecutive figures.
15	*Card playing?*	Have not played for many years, but I am sure should not remember.
16	*Chess?*	Never played.

[In 1880 he published a short paper in 'Nature' (vol. xxi. p. 207) on the "Fertility of Hybrids from the common and Chinese goose." He received the hybrids from the Rev. Dr. Goodacre, and was glad of the opportunity of testing the accuracy of the statement that these species are fertile *inter se.* This fact, which was given in the 'Origin' on the authority of Mr. Eyton, he considered the most remarkable as yet recorded with respect to the fertility of hybrids. The fact (as confirmed by himself and Dr. Goodacre) is of interest as giving another proof that sterility is no criterion of specific difference, since the two species of goose now shown to be fertile *inter se* are so distinct that they have been placed by some authorities in distinct genera or subgenera.

The following letter refers to Mr. Huxley's lecture: "The Coming of Age of the Origin of Species,"* given at the Royal Institution, April 9, 1880, published in 'Nature,' and in 'Science and Culture,' p. 310:]

C. Darwin to T. H. Huxley.

Abinger Hall, Dorking, Sunday, April 11, 1880.

MY DEAR HUXLEY,—I wished much to attend your Lecture, but I have had a bad cough, and we have come here to see whether a change would do me good, as it has done. What a magnificent success your lecture seems to

* This same "Coming of Age" was the subject of an address from the Council of the Otago Institute. It is given in 'Nature,' February 24, 1881.

have been, as I judge from the reports in the *Standard* and *Daily News*, and more especially from the accounts given me by three of my children. I suppose that you have not written out your lecture, so I fear there is no chance of its being printed *in extenso*. You appear to have piled, as on so many other occasions, honours high and thick on my old head. But I well know how great a part you have played in establishing and spreading the belief in the descent-theory, ever since that grand review in the *Times* and the battle royal at Oxford up to the present day.

Ever, my dear Huxley,

Yours sincerely and gratefully,

CHARLES DARWIN.

P.S.—It was absurdly stupid in me, but I had read the announcement of your Lecture, and thought that you meant the maturity of the subject, until my wife one day remarked, "it is almost twenty-one years since the 'Origin' appeared," and then for the first time the meaning of your words flashed on me!

[In the above-mentioned lecture Mr. Huxley made a strong point of the accumulation of palæontological evidence which the years between 1859 and 1880 have given us in favour of Evolution. On this subject my father wrote (August 31, 1880):]

MY DEAR PROFESSOR MARSH,—I received some time ago your very kind note of July 28th, and yesterday the magnificent volume.* I have looked with renewed admiration at the plates, and will soon read the text. Your work on these old birds, and on the many fossil animals of North America, has afforded the best support to the theory of Evolution,

* Odontornithes. A monograph on the extinct Toothed Birds of N. America. 1880. By O. C. Marsh.

which has appeared within the last twenty years.* The general appearance of the copy which you have sent me is worthy of its contents, and I can say nothing stronger than this.

With cordial thanks, believe me,

Yours very sincerely,

CHARLES DARWIN.

[In November, 1880, he received an account of a flood in Brazil, from which his friend Fritz Müller had barely escaped with his life. My father immediately wrote to Hermann Müller anxiously enquiring whether his brother had lost books, instruments, &c., by this accident, and begging in that case "for the sake of science, so that science should not suffer," to be allowed to help in making good the loss. Fortunately, however, the injury to Fitz Müller's possessions was not so great as was expected, and the incident remains only as a memento, which I trust cannot be otherwise than pleasing to the survivor, of the friendship of the two naturalists.

In 'Nature' (November 11, 1880) appeared a letter from my father, which is, I believe, the only instance in which he wrote publicly with anything like severity. The late Sir Wyville Thomson wrote, in the Introduction to the 'Voyage of the *Challenger*': "The character of the abyssal fauna refuses to give the least support to the theory which refers the evolution of species to extreme variation guided only by natural selection." My father, after characterising these remarks as a "standard of criticism, not uncommonly reached by theologians and metaphysicians," goes on to take

* Mr. Huxley has well pointed out ('Science and Culture,' p. 317) that: "In 1875, the discovery of the toothed birds of the cretaceous formation in N. America, by Prof. Marsh, completed the series of transitional forms between birds and reptiles, and removed Mr. Darwin's proposition that, 'many animal forms of life have been utterly lost, through which the early progenitors of birds were formerly connected with the early progenitors of the other vertebrate classes,' from the region of hypothesis to that of demonstrable fact."

exception to the term "extreme variation," and challenges Sir Wyville to name any one who has "said that the evolution of species depends only on natural selection." The letter closes with an imaginary scene between Sir Wyville and a breeder, in which Sir Wyville criticises artificial selection in a somewhat similar manner. The breeder is silent, but on the departure of his critic he is supposed to make use of "emphatic but irreverent language about naturalists." The letter, as originally written, ended with a quotation from Sedgwick on the invulnerability of those who write on what they do not understand, but this was omitted on the advice of a friend, and curiously enough a friend whose combativeness in the good cause my father had occasionally curbed.]

C. Darwin to G. J. Romanes.

Down, April 16, 1881.

MY DEAR ROMANES,—My MS. on 'Worms' has been sent to the printers, so I am going to amuse myself by scribbling to you on a few points; but you must not waste your time in answering at any length this scribble.

Firstly, your letter on intelligence was very useful to me and I tore up and re-wrote what I sent to you. I have not attempted to define intelligence; but have quoted your remarks on experience, and have shown how far they apply to worms. It seems to me that they must be said to work with some intelligence, anyhow they are not guided by a blind instinct.

Secondly, I was greatly interested by the abstract in 'Nature' of your work on Echinoderms,* the complexity with simplicity, and with such curious co-ordination of the nervous system is marvellous; and you showed me before what splendid gymnastic feats they can perform.

* "On the locomotor system of Echinoderms," by G. J. Romanes and J. Cossar Ewart. 'Philosophical Transactions,' 1881, p. 829.

Thirdly, Dr. Roux has sent me a book just published by him: 'Der Kampf der Theile,' &c., 1881 (240 pages in length).

He is manifestly a well-read physiologist and pathologist, and from his position a good anatomist. It is full of reasoning, and this in German is very difficult to me, so that I have only skimmed through each page; here and there reading with a little more care. As far as I can imperfectly judge, it is the most important book on Evolution which has appeared for some time. I believe that G. H. Lewes hinted at the same fundamental idea, viz. that there is a struggle going on within every organism between the organic molecules, the cells and the organs. I think that his basis is, that every cell which best performs its function is, in consequence, at the same time best nourished and best propagates its kind. The book does not touch on mental phenomena, but there is much discussion on rudimentary or atrophied parts, to which subject you formerly attended. Now if you would like to read this book, I would send it. . . . If you read it, and are struck with it (but I may be *wholly* mistaken about its value), you would do a public service by analysing and criticising it in 'Nature.'

Dr. Roux makes, I think, a gigantic oversight in never considering plants; these would simplify the problem for him.

Fourthly, I do not know whether you will discuss in your book on the mind of animals any of the more complex and wonderful instincts. It is unsatisfactory work, as there can be no fossilised instincts, and the sole guide is their state in other members of the same order, and mere *probability*.

But if you do discuss any (and it will perhaps be expected of you), I should think that you could not select a better case than that of the sand wasps, which paralyse their prey, as formerly described by Fabre, in his wonderful paper in the 'Annales des Sciences,' and since amplified in his admirable 'Souvenirs.'

Whilst reading this latter book, I speculated a little on the subject. Astonishing nonsense is often spoken of the sand wasp's knowledge of anatomy. Now will any one say that the Gauchos on the plains of La Plata have such knowledge, yet I have often seen them pith a struggling and lassoed cow on the ground with unerring skill, which no mere anatomist could imitate. The pointed knife was infallibly driven in between the vertebræ by a single slight thrust. I presume that the art was first discovered by chance, and that each young Gaucho sees exactly how the others do it, and then with a very little practice learns the art. Now I suppose that the sand wasps originally merely killed their prey by stinging them in many places (see p. 129 of Fabre's 'Souvenirs,' and p. 241) on the lower and softest side of the body—and that to sting a certain segment was found by far the most successful method; and was inherited like the tendency of a bulldog to pin the nose of a bull, or of a ferret to bite the cerebellum. It would not be a very great step in advance to prick the ganglion of its prey only slightly, and thus to give its larvæ fresh meat instead of old dried meat. Though Fabre insists so strongly on the unvarying character of instinct, yet it is shown that there is some variability, as at p. 176, 177.

I fear that I shall have utterly wearied you with my scribbling and bad handwriting.

My dear Romanes, yours very sincerely,

CH. DARWIN.

Postscript of a Letter to Professor A. Agassiz, May 5th, 1881 :—

"I read with much interest your address before the American Association. However true your remarks on the genealogies of the several groups may be, I hope and believe that you have over-estimated the difficulties to be encountered in the future :—A few days after reading your address, I interpreted to myself your remarks on one point (I hope in some degree correctly) in the following fashion :—

Any character of an ancient, generalised, or intermediate form may, and often does, re-appear in its descendants, after countless generations, and this explains the extraordinarily complicated affinities of existing groups. This idea seems to me to throw a flood of light on the lines, sometimes used to represent affinities, which radiate in all directions, often to very distant sub-groups,—a difficulty which has haunted me for half a century. A strong case could be made out in favour of believing in such reversion after immense intervals of time. I wish the idea had been put into my head in old days, for I shall never again write on difficult subjects, as I have seen too many cases of old men becoming feeble in their minds, without being in the least conscious of it. If I have interpreted your ideas at all correctly, I hope that you will re-urge, on any fitting occasion, your view. I have mentioned it to a few persons capable of judging, and it seemed quite new to them. I beg you to forgive the proverbial garrulity of old age.

C. D."

[The following letter refers to Sir J. D. Hooker's Geographical address at the York Meeting (1881) of the British Association :]

C. Darwin to J. D. Hooker.

Down, August 6, 1881.

MY DEAR HOOKER,—For Heaven's sake never speak of boring me, as it would be the greatest pleasure to aid you in the slightest degree and your letter has interested me exceedingly. I will go through your points seriatim, but I have never attended much to the history of any subject, and my memory has become atrociously bad. It will therefore be a mere chance whether any of my remarks are of any use.

Your idea, to show what travellers have done, seems to me a brilliant and just one, especially considering your audience.

1. I know nothing about Tournefort's works.

2. I believe that you are fully right in calling Humboldt the greatest scientific traveller who ever lived. I have lately read two or three volumes again. His Geology is funny stuff; but that merely means that he was not in advance of his age. I should say he was wonderful, more for his near approach to omniscience than for originality. Whether or not his position as a scientific man is as eminent as we think, you might truly call him the parent of a grand progeny of scientific travellers, who, taken together, have done much for science.

3. It seems to me quite just to give Lyell (and secondarily E. Forbes) a very prominent place.

4. Dana was, I believe, the first man who maintained the permanence of continents and the great oceans. . . . When I read the '*Challenger's*' conclusion that sediment from the land is not deposited at greater distances than 200 to 300 miles from the land, I was much strengthened in my old belief. Wallace seems to me to have argued the case excellently. Nevertheless, I would speak, if I were in your place, rather cautiously; for T. Mellard Reade has argued lately with some force against the view; but I cannot call to mind his arguments. If forced to express a judgment, I should abide by the view of approximate permanence since Cambrian days.

5. The extreme importance of the Arctic fossil plants, is self-evident. Take the opportunity of groaning over [our] ignorance of the Lignite Plants of Kerguelen Land, or any Antarctic land. It might do good.

6. I cannot avoid feeling sceptical about the travelling of plants from the North *except during the Tertiary period.* It may of course have been so and probably was so from one of the two poles at the earliest period, during Pre-Cambrian ages; but such speculations seem to me hardly scientific, seeing how little we know of the old Floras.

I will now jot down without any order a few miscellaneous remarks.

I think you ought to allude to Alph. De Candolle's great book, for though it (like almost everything else) is washed out of my mind, yet I remember most distinctly thinking it a very valuable work. Anyhow, you might allude to his excellent account of the history of all cultivated plants.

How shall you manage to allude to your New Zealand and Tierra del Fuego work? if you do not allude to them you will be scandalously unjust.

The many Angiosperm plants in the Cretacean beds of the United States (and as far as I can judge the age of these beds has been fairly well made out) seems to me a fact of very great importance, so is their relation to the existing flora of the United States under an Evolutionary point of view. Have not some Australian extinct forms been lately found in Australia? or have I dreamed it?

Again, the recent discovery of plants rather low down in our Silurian beds is very important.

Nothing is more extraordinary in the history of the Vegetable Kingdom, as it seems to me, than the *apparently* very sudden or abrupt development of the higher plants. I have sometimes speculated whether there did not exist somewhere during long ages an extremely isolated continent, perhaps near the South Pole.

Hence I was greatly interested by a view which Saporta propounded to me, a few years ago, at great length in MS. and which I fancy he has since published, as I urged him to do—viz., that as soon as flower-frequenting insects were developed, during the latter part of the secondary period, an enormous impulse was given to the development of the higher plants by cross-fertilization being thus suddenly formed.

A few years ago I was much struck with Axel Blytt's * Essay showing from observation, on the peat beds in Scandi-

* See footnote, Vol. iii. p. 215.

navia, that there had apparently been long periods with more rain and other with less rain (perhaps connected with Croll's recurrent astronomical periods), and that these periods had largely determined the present distribution of the plants of Norway and Sweden. This seemed to me a very important essay.

I have just read over my remarks and I fear that they will not be of the slightest use to you.

I cannot but think that you have got through the hardest, or at least the most difficult, part of your work in having made so good and striking a sketch of what you intend to say; but I can quite understand how you must groan over the great necessary labour.

I most heartily sympathise with you on the successes of B. and R.: as years advance what happens to oneself becomes of very little consequence, in comparison with the careers of our children.

Keep your spirits up, for I am convinced that you will make an excellent address.

Ever yours affectionately,
CHARLES DARWIN.

[In September he wrote :—

"I have this minute finished reading your splendid but too short address. I cannot doubt that it will have been fully appreciated by the Geographers at York; if not, they are asses and fools."]

C. Darwin to John Lubbock.

Sunday evening [1881].

MY DEAR L.,—Your address * has made me think over what have been the great steps in Geology during the last fifty years, and there can be no harm in telling you my impression. But it is very odd that I cannot remember what

* Presidential Address at the York Meeting of the British Association.

you have said on Geology. I suppose that the classification of the Silurian and Cambrian formations must be considered the greatest or most important step; for I well remember when all these older rocks were called grau-wacke, and nobody dreamed of classing them; and now we have three azoic formations pretty well made out beneath the Cambrian! But the most striking step has been the discovery of the Glacial period: you are too young to remember the prodigious effect this produced about the year 1840 (?) on all our minds. Elie de Beaumont never believed in it to the day of his death! the study of the glacial deposits led to the study of the superficial drift, which was formerly *never studied* and called Diluvium, as I well remember. The study under the microscope of rock-sections is another not inconsiderable step. So again the making out of cleavage and the foliation of the metamorphic rocks. But I will not run on, having now eased my mind. Pray do not waste even one minute in acknowledging my horrid scrawls.

Ever yours,
CH. DARWIN.

[The following extracts referring to the late Francis Maitland Balfour,* show my father's estimate of his work and intellectual qualities, but they give merely an indication of his strong appreciation of Balfour's most loveable personal character:—

From a letter to Fritz Müller, January 5, 1882:—

"Your appreciation of Balfour's book ['Comparative Embryology'] has pleased me excessively, for though I could not properly judge of it, yet it seemed to me one of the most remarkable books which have been published for some considerable time. He is quite a young man, and if he keeps

* Professor of Animal Morphology at Cambridge. He was born 1851, and was killed, with his guide, on the Aiguille Blanche, near Courmayeur, in July, 1882.

his health, will do splendid work. . . . He has a fair fortune of his own, so that he can give up his whole time to Biology. He is very modest, and very pleasant, and often visits here and we like him very much."

From a letter to Dr. Dohrn, February 13, 1882:—

"I have got one very bad piece of news to tell you, that F. Balfour is very ill at Cambridge with typhoid fever.... I hope that he is not in a very dangerous state; but the fever is severe. Good Heavens, what a loss he would be to Science, and to his many loving friends!"]

C. Darwin to T. H. Huxley.

Down, January 12, 1882.

MY DEAR HUXLEY,—Very many thanks for 'Science and Culture,' and I am sure that I shall read most of the essays with much interest. With respect to Automatism,* I wish that you could review yourself in the old, and of course forgotten, trenchant style, and then you would here answer yourself with equal incisiveness; and thus, by Jove, you might go on *ad infinitum*, to the joy and instruction of the world.

Ever yours very sincerely,
CHARLES DARWIN.

[The following letter refers to Dr. Ogle's translation of Aristotle, 'On the Parts of Animals' (1882):]

C. Darwin to W. Ogle.

Down, February 22, 1882.

MY DEAR DR. OGLE,—You must let me thank you for the pleasure which the introduction to the Aristotle book

* "On the hypothesis that animals are automata and its history," an Address given at the Belfast meeting of the British Association, 1874, and published in the 'Fortnightly Review,' 1874, and in 'Science and Culture.'

has given me. I have rarely read anything which has interested me more, though I have not read as yet more than a quarter of the book proper.

From quotations which I had seen, I had a high notion of Aristotle's merits, but I had not the most remote notion what a wonderful man he was. Linnæus and Cuvier have been my two gods, though in very different ways, but they were mere schoolboys to old Aristotle. How very curious, also, his ignorance on some points, as on muscles as the means of movement. I am glad that you have explained in so probable a manner some of the grossest mistakes attributed to him. I never realized, before reading your book, to what an enormous summation of labour we owe even our common knowledge. I wish old Aristotle could know what a grand Defender of the Faith he had found in you. Believe me, my dear Dr. Ogle,

Yours very sincerely,

CH. DARWIN.

[In February, he received a letter and a specimen from a Mr. W. D. Crick, which illustrated a curious mode of dispersal of bivalve shells, namely, by closure of their valves so as to hold on to the leg of a water-beetle. This class of fact had a special charm for him, and he wrote to 'Nature' describing the case.*

In April, he received a letter from Dr. W. Van Dyck, Lecturer in Zoology at the Protestant College of Beyrout. The letter showed that the street dogs of Beyrout had been rapidly mongrelised by introduced European dogs, and the facts have an interesting bearing on my father's theory of Sexual Selection.]

* 'Nature,' April 6, 1882.

C. Darwin to W. Van Dyck.

Down, April 3, 1882.

DEAR SIR,—After much deliberation, I have thought it best to send your very interesting paper to the Zoological Society, in hopes that it will be published in their Journal. This journal goes to every scientific institution in the world, and the contents are abstracted in all year-books on Zoology. Therefore I have preferred it to 'Nature,' though the latter has a wider circulation, but is ephemeral.

I have prefaced your essay by a few general remarks, to which I hope that you will not object.

Of course I do not know that the Zoological Society, which is much addicted to mere systematic work, will publish your essay. If it does, I will send you copies of your essay, but these will not be ready for some months. If not published by the Zoological Society, I will endeavour to get 'Nature' to publish it. I am very anxious that it should be published and preserved. Dear Sir,

Yours faithfully,

CH. DARWIN.

[The paper was read at a meeting of the Zoological Society on April 18th—the day before my father's death.

The preliminary remarks with which Dr. Van Dyck's paper is prefaced are thus the latest of my father's writings.]

We must now return to an early period of his life, and give a connected account of his botanical work, which has hitherto been omitted.

CHAPTER VII.

FERTILISATION OF FLOWERS.

[IN the letters already given we have had occasion to notice the general bearing of a number of botanical problems on the wider question of Evolution. The detailed work in botany which my father accomplished by the guidance of the light cast on the study of natural history by his own work on Evolution remains to be noticed. In a letter to Mr. Murray, September 24th, 1861, speaking of his book on the 'Fertilisation of Orchids,' he says: "It will perhaps serve to illustrate how Natural History may be worked under the belief of the modification of species." This remark gives a suggestion as to the value and interest of his botanical work, and it might be expressed in far more emphatic language without danger of exaggeration.

In the same letter to Mr. Murray, he says: "I think this little volume will do good to the 'Origin,' as it will show that I have worked hard at details." It is true that his botanical work added a mass of corroborative detail to the case for Evolution, but the chief support to his doctrines given by these researches was of another kind. They supplied an argument against those critics who have so freely dogmatised as to the uselessness of particular structures, and as to the consequent impossibility of their having been developed by means of natural selection. His observations on Orchids enabled him to say: "I can show the meaning of some of the apparently meaningless ridges, horns; who will now venture to say that this or that structure is useless?" A kindred point is expressed in a letter to Sir J. D. Hooker (May 14th, 1862):—

"When many parts of structure, as in the woodpecker, show distinct adaptation to external bodies, it is preposterous to attribute them to the effects of climate, &c., but when a single point alone, as a hooked seed, it is conceivable it may thus have arisen. I have found the study of Orchids eminently useful in showing me how nearly all parts of the flower are co-adapted for fertilisation by insects, and therefore the results of natural selection,—even the most trifling details of structure."

One of the greatest services rendered by my father to the study of Natural History is the revival of Teleology. The evolutionist studies the purpose or meaning of organs with the zeal of the older Teleology, but with far wider and more coherent purpose. He has the invigorating knowledge that he is gaining not isolated conceptions of the economy of the present, but a coherent view of both past and present. And even where he fails to discover the use of any part, he may, by a knowledge of its structure, unravel the history of the past vicissitudes in the life of the species. In this way a vigour and unity is given to the study of the forms of organised beings, which before it lacked. This point has already been discussed in Mr. Huxley's chapter on the 'Reception of the *Origin of Species*,' and need not be here considered. It does, however, concern us to recognize that this "great service to natural science," as Dr. Gray describes it, was effected almost as much by his special botanical work as by the 'Origin of Species.'

For a statement of the scope and influence of my father's botanical work, I may refer to Mr. Thiselton Dyer's article in 'Charles Darwin,' one of the *Nature Series*. Mr. Dyer's wide knowledge, his friendship with my father, and especially his power of sympathising with the work of others, combine to give this essay a permanent value. The following passage (p. 43) gives a true picture :—

"Notwithstanding the extent and variety of his botanical work, Mr. Darwin always disclaimed any right to be regarded as a professed botanist. He turned his attention to plants, doubtless because they were convenient objects for studying organic phenomena in their least complicated forms; and this point of view, which, if one may use the expression without disrespect, had something of the amateur about it, was in itself of the greatest importance. For, from not being, till he took up any point, familiar with the literature bearing on it, his mind was absolutely free from any prepossession. He was never afraid of his facts, or of framing any hypothesis, however startling, which seemed to explain them. . . . In any one else such an attitude would have produced much work that was crude and rash. But Mr. Darwin—if one may venture on language which will strike no one who had conversed with him as over-strained—seemed by gentle persuasion to have penetrated that reserve of nature which baffles smaller men. In other words, his long experience had given him a kind of instinctive insight into the method of attack of any biological problem, however unfamiliar to him, while he rigidly controlled the fertility of his mind in hypothetical explanations by the no less fertility of ingeniously devised experiment."

To form any just idea of the greatness of the revolution worked by my father's researches in the study of the fertilisation of flowers, it is necessary to know from what a condition this branch of knowledge has emerged. It should be remembered that it was only during the early years of the present century that the idea of sex, as applied to plants, became firmly established. Sachs, in his 'History of Botany' (1875), has given some striking illustrations of the remarkable slowness with which its acceptance gained ground. He remarks that when we consider the experimental proofs given by Camerarius (1694), and by Kölreuter (1761–66), it appears incredible that doubts should afterwards have been raised as to the sexuality of plants. Yet he shows that such doubts did actually repeatedly crop up. These adverse criticisms rested for the most part on careless experiments, but in many cases on *à priori* arguments. Even as late as 1820, a book of this kind, which would now rank with circle squaring, or flat-earth philosophy, was seriously noticed in a botanical journal.

A distinct conception of sex as applied to plants had not long emerged from the mists of profitless discussion and feeble experiment, at the time when my father began botany by attending Henslow's lectures at Cambridge.

When the belief in the sexuality of plants had become established as an incontrovertible piece of knowledge, a weight of misconception remained, weighing down any rational view of the subject. Camerarius * believed (naturally enough in his day) that hermaphrodite flowers are necessarily self-fertilised. He had the wit to be astonished at this, a degree of intelligence which, as Sachs points out, the majority of his successors did not attain to.

The following extracts from a note-book show that this point occurred to my father as early as 1837:—

"Do not plants which have male and female organs together [*i.e.* in the same flower] yet receive influence from other plants? Does not Lyell give some argument about varieties being difficult to keep [true] on account of pollen from other plants? Because this may be applied to show all plants do receive intermixture."

Sprengel, † indeed, understood that the hermaphrodite structure of flowers by no means necessarily leads to self-fertilisation. But although he discovered that in many cases pollen is of necessity carried to the stigma of another *flower*, he did not understand that in the advantage gained by the

* Sachs, 'Geschichte,' p. 419.

† Christian Conrad Sprengel, born 1750, died 1816.

intercrossing of distinct *plants* lies the key to the whole question. Hermann Müller has well remarked that this "omission was for several generations fatal to Sprengel's work. . . . . For both at the time and subsequently, botanists felt above all the weakness of his theory, and they set aside, along with his defective ideas, his rich store of patient and acute observations and his comprehensive and accurate interpretations." It remained for my father to convince the world that the meaning hidden in the structure of flowers was to be found by seeking light in the same direction in which Sprengel, seventy years before, had laboured. Robert Brown was the connecting link between them; for although, according to Dr. Gray,* Brown, in common with the rest of the world, looked on Sprengel's ideas as fantastic, yet it was at his recommendation that my father in 1841 read Sprengel's now celebrated 'Secret of Nature Displayed.† The book impressed him as being "full of truth," although "with some little nonsense." It not only encouraged him in kindred speculation, but guided him in his work, for in 1844 he speaks of verifying Sprengel's observations. It may be doubted whether Robert Brown ever planted a more fruitful seed than in putting such a book into such hands.

A passage in the 'Autobiography' (vol. i. p. 90) shows how it was that my father was attracted to the subject of fertilisation: "During the summer of 1839, and I believe during the previous summer, I was led to attend to the cross-fertilisation of flowers by the aid of insects, from having come to the conclusion in my speculations on the origin of species, that crossing played an important part in keeping specific forms constant."

The original connection between the study of flowers and the problem of Evolution is curious, and could hardly have been predicted. Moreover, it was not a permanent bond.

* 'Nature,' 1874, p. 80.

† 'Das entdeckte Geheimniss der Natur im Baue und in der Befruchtung der Blumen.' Berlin, 1793.

As soon as the idea arose that the offspring of cross-fertilisation is, in the struggle for life, likely to conquer the seedlings of self-fertilised parentage, a far more vigorous belief in the potency of natural selection in moulding the structure of flowers is attained. A central idea is gained towards which experiment and observation may be directed.

Dr. Gray has well remarked with regard to this central idea ('Nature,' June 4, 1874):—"The aphorism, 'Nature abhors a vacuum,' is a characteristic specimen of the science of the middle ages. The aphorism, 'Nature abhors close fertilisation,' and the demonstration of the principle, belong to our age and to Mr. Darwin. To have originated this, and also the principle of Natural Selection . . . . and to have applied these principles to the system of nature, in such a manner as to make, within a dozen years, a deeper impression upon natural history than has been made since Linnæus, is ample title for one man's fame."

The flowers of the Papilionaceæ attracted his attention early, and were the subject of his first paper on fertilisation.* The following extract from an undated letter to Dr. Asa Gray seems to have been written before the publication of this paper, probably in 1856 or 1857:—

". . . . What you say on Papilionaceous flowers is very true; and I have no facts to show that varieties are crossed; but yet (and the same remark is applicable in a beautiful way to Fumaria and Dielytra, as I noticed many years ago), I must believe that the flowers are constructed partly in direct relation to the visits of insects; and how insects can avoid bringing pollen from other individuals I cannot understand. It is really pretty to watch the action of a Humble-bee on the scarlet kidney bean, and in this genus (and in *Lathyrus*

* *Gardeners' Chronicle*, 1857, p. 725. It appears that this paper was a piece of "over-time" work. He wrote to a friend, "that confounded leguminous paper was done in the afternoon, and the consequence was I had to go to Moor Park for a week."

grandiflorus) the honey is so placed that the bee invariably alights on that *one* side of the flower towards which the spiral pistil is protruded (bringing out with it pollen), and by the depression of the wing-petal is forced against the bee's side all dusted with pollen.* In the broom the pistil is rubbed on the centre of the back of the bee. I suspect there is something to be made out about the Leguminosæ, which will bring the case within *our* theory; though I have failed to do so. Our theory will explain why in the vegetable and animal kingdom the act of fertilisation even in hermaphrodites usually takes place sub-jove, though thus exposed to *great* injury from damp and rain. In animals which cannot be [fertilised] by insects or wind, there is *no case* of *land*-animals being hermaphrodite without the concourse of two individuals."

A letter to Dr. Asa Gray (Sept. 5th, 1857) gives the substance of the paper in the *Gardeners' Chronicle*:—

"Lately I was led to examine buds of kidney bean with the pollen shed; but I was led to believe that the pollen could *hardly* get on the stigma by wind or otherwise, except by bees visiting [the flower] and moving the wing petals: hence I included a small bunch of flowers in two bottles in every way treated the same: the flowers in one I daily just momentarily moved, as if by a bee; these set three fine pods, the other *not one*. Of course this little experiment must be tried again, and this year in England it is too late, as the flowers seem now seldom to set. If bees are necessary to this flower's self-fertilisation, bees must almost cross them, as their dusted right-side of head and right legs constantly touch the stigma.

"I have, also, lately been re-observing daily *Lobelia fulgens*—this in my garden is never visited by insects, and never sets

* If you will look at a bed of scarlet kidney beans you will find that the wing-petals on the *left* side alone are all scratched by the tarsi of the bees. [Note in the original letter by C. Darwin.]

seeds, without pollen be put on the stigma (whereas the small blue Lobelia is visited by bees and does set seed); I mention this because there are such beautiful contrivances to prevent the stigma ever getting its own pollen; which seems only explicable on the doctrine of the advantage of crosses."

The paper was supplemented by a second in 1858.* The chief object of these publications seems to have been to obtain information as to the possibility of growing varieties of leguminous plants near each other, and yet keeping them true. It is curious that the Papilionaceæ should not only have been the first flowers which attracted his attention by their obvious adaptation to the visits of insects, but should also have constituted one of his sorest puzzles. The common pea and the sweet pea gave him much difficulty, because, although they are as obviously fitted for insect-visits as the rest of the order, yet their varieties keep true. The fact is that neither of these plants being indigenous, they are not perfectly adapted for fertilisation by British insects. He could not, at this stage of his observations, know that the co-ordination between a flower and the particular insect which fertilises it may be as delicate as that between a lock and its key, so that this explanation was not likely to occur to him.†

Besides observing the Leguminosæ, he had already begun, as shown in the foregoing extracts, to attend to the structure of other flowers in relation to insects. At the beginning of 1860 he worked at Leschenaultia,‡ which at first puzzled him,

* *Gardeners' Chronicle*, 1858, p. 828. In 1861 another paper on Fertilisation appeared in the *Gardeners' Chronicle*, p. 552, in which he explained the action of insects on *Vinca major*. He was attracted to the periwinkle by the fact that it is not visited by insects and never sets seeds.

† He was of course alive to variety in the habits of insects. He published a short note in the *Entomologist's Weekly Intelligencer*, 1860, asking whether the Tineina and other small moths suck flowers.

‡ He published a short paper on the manner of fertilisation of this flower, in the *Gardeners' Chronicle*, 1871, p. 1166.

but was ultimately made out. A passage in a letter chiefly relating to Leschenaultia seems to show that it was only in the spring of 1860 that he began widely to apply his knowledge to the relation of insects to other flowers. This is somewhat surprising, when we remember that he had read Sprengel many years before. He wrote (May 14):—

"I should look at this curious contrivance as specially related to visits of insects; as I begin to think is almost universally the case."

Even in July 1862 he wrote to Dr. Asa Gray:—

"There is no end to the adaptations. Ought not these cases to make one very cautious when one doubts about the use of all parts? I fully believe that the structure of all irregular flowers is governed in relation to insects. Insects are the Lords of the floral (to quote the witty *Athenæum*) world."

He was probably attracted to the study of Orchids by the fact that several kinds are common near Down. The letters of 1860 show that these plants occupied a good deal of his attention; and in 1861 he gave part of the summer and all the autumn to the subject. He evidently considered himself idle for wasting time on Orchids which ought to have been given to 'Variation under Domestication.' Thus he wrote:—

"There is to me incomparably more interest in observing than in writing; but I feel quite guilty in trespassing on these subjects, and not sticking to varieties of the confounded cocks, hens and ducks. I hear that Lyell is savage at me. I shall never resist Linum next summer."

It was in the summer of 1860 that he made out one of the most striking and familiar facts in the book, namely, the manner in which the pollen masses in Orchis are adapted for removal by insects. He wrote to Sir J. D. Hooker July 12:—

"I have been examining *Orchis pyramidalis*, and it almost

equals, perhaps even beats, your Listera case; the sticky glands are congenitally united into a saddle-shaped organ, which has great power of movement, and seizes hold of a bristle (or proboscis) in an admirable manner, and then another movement takes place in the pollen masses, by which they are beautifully adapted to leave pollen on the two *lateral* stigmatic surfaces. I never saw anything so beautiful."

In June of the same year he wrote:—

"You speak of adaptation being rarely *visible*, though present in plants. I have just recently been looking at the common Orchis, and I declare I think its adaptations in every part of the flower quite as beautiful and plain, or even more beautiful than in the Woodpecker. I have written and sent a notice for the *Gardeners' Chronicle*,* on a curious difficulty in the Bee Orchis, and should much like to hear what you think of the case. In this article I have incidentally touched on adaptation to visits of insects; but the contrivance to keep the sticky glands fresh and sticky beats almost everything in nature. I never remember having seen it described, but it must have been, and, as I ought not in my book to give the observation as my own, I should be very glad to know where this beautiful contrivance is described."

He wrote also to Dr. Gray, June 8, 1860:—

"Talking of adaptation, I have lately been looking at our common orchids, and I dare say the facts are as old and well-known as the hills, but I have been so struck with admiration at the contrivances, that I have sent a notice to the *Gardeners' Chronicle.* The *Ophrys apifera,* offers, as you will see, a curious contradiction in structure."

Besides attending to the fertilisation of the flowers he was already, in 1860, busy with the homologies of the parts, a

* June 9, 1860. This seems to have attracted some attention, especially among entomologists, as it was reprinted in the *Entomologist's Weekly Intelligencer*, 1860.

subject of which he made good use in the Orchid book. He wrote to Sir Joseph Hooker (July):—

"It is a real good joke my discussing homologies of Orchids with you, after examining only three or four genera; and this very fact makes me feel positive I am right!! I do not quite understand some of your terms; but sometime I must get you to explain the homologies; for I am intensely interested on the subject, just as at a game of chess."

This work was valuable from a systematic point of view. In 1880 he wrote to Mr. Bentham:—

"It was very kind in you to write to me about the Orchideæ, for it has pleased me to an extreme degree that I could have been of the *least* use to you about the nature of the parts."

The pleasure which his early observations on Orchids gave him is shown in such extracts as the following from a letter to Sir J. D. Hooker (July 27, 1861):—

"You cannot conceive how the Orchids have delighted me. They came safe, but box rather smashed; cylindrical old cocoa- or snuff-canister much safer. I enclose postage. As an account of the movement, I shall allude to what I suppose is Oncidium, to make *certain*,—is the enclosed flower with crumpled petals this genus? Also I most specially want to know what the enclosed little globular brown Orchid is. I have only seen pollen of a Cattleya on a bee, but surely have you not unintentionally sent me what I wanted most (after Catasetum or Mormodes), viz. one of the Epidendreæ?! I *particularly* want (and will presently tell you why) another spike of this little Orchid, with older flowers, some even almost withered."

His delight in observation is again shown in a letter to Dr. Gray (1863). Referring to Crüger's letters from Trinidad, he wrote:—"Happy man, he has actually seen crowds of bees flying round Catasetum, with the pollinia sticking to their backs!"

The following extracts of letters to Sir J. D. Hooker illustrate further the interest which his work excited in him:—

"Veitch sent me a grand lot this morning. What wonderful structures!

"I have now seen enough, and you must not send me more, for though I enjoy looking at them *much*, and it has been very useful to me, seeing so many different forms, it is idleness. For my object each species requires studying for days. I wish you had time to take up the group. I would give a good deal to know what the rostellum is, of which I have traced so many curious modifications. I suppose it cannot be one of the stigmas,* there seems a great tendency for two lateral stigmas to appear. My paper, though touching on only subordinate points will run, I fear, to 100 MS. folio pages! The beauty of the adaptation of parts seems to me unparalleled. I should think or guess waxy pollen was most differentiated. In Cypripedium which seems least modified, and a much exterminated group, the grains are single. In *all others*, as far as I have seen, they are in packets of four; and these packets cohere into many wedge-formed masses in Orchis; into eight, four, and finally two. It seems curious that a flower should exist, which could *at most* fertilise only two other flowers, seeing how abundant pollen generally is; this fact I look at as explaining the perfection of the contrivance by which the pollen, so important from its fewness, is carried from flower to flower" (1861).

"I was thinking of writing to you to-day, when your note with the Orchids came. What frightful trouble you have taken about Vanilla; you really must not take an atom more; for the Orchids are more play than real work. I have been much interested by Epidendrum, and have worked all morning at them; for heaven's sake, do not corrupt me by any more" (August 30, 1861).

* It is a modification of the upper stigma.

He originally intended to publish his notes on Orchids as a paper in the Linnean Society's Journal, but it soon became evident that a separate volume would be a more suitable form of publication. In a letter to Sir J. D. Hooker, Sept. 24, 1861, he writes:—

"I have been acting, I fear that you will think, like a goose; and perhaps in truth I have. When I finished a few days ago my Orchis paper, which turns out 140 folio pages!! and thought of the expense of woodcuts, I said to myself, I will offer the Linnean Society to withdraw it, and publish it in a pamphlet. It then flashed on me that perhaps Murray would publish it, so I gave him a cautious description, and offered to share risks and profits. This morning he writes that he will publish and take all risks, and share profits and pay for all illustrations. It is a risk, and heaven knows whether it will not be a dead failure, but I have not deceived Murray, and [have] told him that it would interest those alone who cared much for natural history. I hope I do not exaggerate the curiosity of the many special contrivances."

He wrote the two following letters to Mr. Murray about the publication of the book:]

Down, Sept. 21 [1861].

MY DEAR SIR,—Will you have the kindness to give me your opinion, which I shall implicitly follow. I have just finished a very long paper intended for Linnean Society (the title is enclosed), and yesterday for the first time it occurred to me that *possibly* it might be worth publishing separately, which would save me trouble and delay. The facts are new, and have been collected during twenty years and strike me as curious. Like a Bridgewater treatise, the chief object is to show the perfection of the many contrivances in Orchids. The subject of propagation is interesting to most people, and is treated in my paper so that any woman could read it. Parts are dry and purely scientific; but I think my paper would interest a good many of such persons who care for Natural History, but no others.

. . . It would be a very little book, and I believe you think very little books objectionable. I have myself *great* doubts on the subject. I am very apt to think that my geese are swans; but the subject seems to me curious and interesting.

I beg you not to be guided in the least in order to oblige me, but as far as you can judge, please give me your opinion. If I were to publish separately, I would agree to any terms, such as half risk and half profit, or what you liked; but I would not publish on my sole risk, for to be frank, I have been told that no publisher whatever, under such circumstances, cares for the success of a book.

C. Darwin to J. Murray.

Down, Sept. 24 [1861].

MY DEAR SIR,—I am very much obliged for your note and very liberal offer. I have had some qualms and fears. All that I can feel sure of is that the MS. contains many new and curious facts, and I am sure the Essay would have interested me, and will interest those who feel lively interest in the wonders of nature; but how far the public will care for such minute details, I cannot at all tell. It is a bold experiment; and at worst, cannot entail much loss; as a certain amount of sale will, I think, be pretty certain. A large sale is out of the question. As far as I can judge, generally the points which interest me I find interest others; but I make the experiment with fear and trembling,—not for my own sake, but for yours. . . .

[On Sept. 28th he wrote to Sir J. D. Hooker:—

"What a good soul you are not to sneer at me, but to pat me on the back. I have the greatest doubt whether I am not going to do, in publishing my paper, a most ridiculous thing.

It would annoy me much, but only for Murray's sake, if the publication were a dead failure."

There was still much work to be done, and in October he was still receiving Orchids from Kew, and wrote to Hooker:—

"It is impossible to thank you enough. I was almost mad at the wealth of Orchids." And again—

"Mr. Veitch most generously has sent me two splendid buds of Mormodes, which will be capital for dissection, but I fear will never be irritable; so for the sake of charity and love of heaven do, I beseech you, observe what movement takes place in Cychnoches, and what part must be touched. Mr. V. has also sent me one splendid flower of Catasetum, the most wonderful Orchid I have seen."

On Oct. 13th he wrote to Sir Joseph Hooker:—

"It seems that I cannot exhaust your good nature. I have had the hardest day's work at Catasetum and buds of Mormodes, and believe I understand at last the mechanism of movements and the functions. Catasetum is a beautiful case of slight modification of structure leading to new functions. I never was more interested in any subject in my life than in this of Orchids. I owe very much to you."

Again to the same friend, Nov. 1, 1861:—

"If you really can spare another Catasetum, when nearly ready, I shall be most grateful; had I not better send for it? The case is truly marvellous; the (so-called) sensation, or stimulus from a light touch is certainly transmitted through the antennæ for more than one inch *instantaneously*. . . . A cursed insect or something let my last flower off last night."

Professor de Candolle has remarked* of my father, "Ce n'est pas lui qui aurait demandé de construire des palais pour y loger des laboratoires." This was singularly true of his orchid work, or rather it would be nearer the truth to say that he had no laboratory, for it was only after the publication

* 'Darwin considéré, &c.,' 'Archives des Sciences Physiques et Naturelles,' 3ème période. Tome vii. 481, 1882 (May).

of the 'Fertilisation of Orchids,' that he built himself a greenhouse. He wrote to Sir J. D. Hooker (Dec. 24th, 1862):—

"And now I am going to tell you a *most* important piece of news!! I have almost resolved to build a small hot-house; my neighbour's really first-rate gardener has suggested it, and offered to make me plans, and see that it is well done, and he is really a clever fellow, who wins lots of prizes, and is very observant. He believes that we should succeed with a little patience; it will be a grand amusement for me to experiment with plants."

Again he wrote (Feb. 15th, 1863):—

"I write now because the new hot-house is ready, and I long to stock it, just like a schoolboy. Could you tell me pretty soon what plants you can give me; and then I shall know what to order? And do advise me how I had better get such plants as you can *spare*. Would it do to send my tax-cart early in the morning, on a day that was not frosty, lining the cart with mats, and arriving here before night? I have no idea whether this degree of exposure (and of course the cart would be cold) could injure stove-plants; they would be about five hours (with bait) on the journey home."

A week later he wrote:—

"You cannot imagine what pleasure your plants give me (far more than your dead Wedgwood ware can give you); H. and I go and gloat over them, but we privately confessed to each other, that if they were not our own, perhaps we should not see such transcendent beauty in each leaf."

And in March, when he was extremely unwell he wrote:—

"A few words about the Stove-plants; they do so amuse me. I have crawled to see them two or three times. Will you correct and answer, and return enclosed. I have hunted in all my books and cannot find these names,* and I like much to know the family."

* His difficulty with regard to the names of plants is illustrated, with regard to a Lupine on which he was at work, in an extract from

The book was published May 15th, 1862. Of its reception he writes to Mr. Murray, June 13th and 18th:—

"The Botanists praise my Orchid-book to the skies. Some one sent me (perhaps you) the 'Parthenon,' with a good review. The *Athenæum* * treats me with very kind pity and contempt; but the reviewer knew nothing of his subject."

"There is a superb, but I fear exaggerated, review in the 'London Review.' † But I have not been a fool, as I thought I was, to publish; ‡ for Asa Gray, about the most competent judge in the world, thinks almost as highly of the book as does the 'London Review.' The *Athenæum* will hinder the sale greatly."

The Rev. M. J. Berkeley was the author of the notice in the 'London Review,' as my father learned from Sir J. D. Hooker, who added, "I thought it very well done indeed. I have read a good deal of the Orchid-book, and echo all he says."

To this my father replied (June 30th, 1862):—

"My dear Old Friend,—You speak of my warming the cockles of your heart, but you will never know how often you have warmed mine. It is not your approbation of my scientific work (though I care for that more than for any one's): it is something deeper. To this day I remember keenly a letter you wrote to me from Oxford, when I was at the Water-cure, and how it cheered me when I was utterly weary of life.

---

a letter (July 21, 1866) to Sir J. D. Hooker: "I sent to the nursery garden, whence I bought the seed, and could only hear that it was 'the common blue Lupine,' the man saying 'he was no scholard, and did not know Latin, and that parties who make experiments ought to find out the names.'"

* May 24, 1862.

† June 14, 1862.

‡ Doubts on this point still, however, occurred to him about this time. He wrote to Prof. Oliver (June 8): "I am glad that you have read my Orchis-book and seem to approve of it; for I never published anything which I so much doubted whether it was worth publishing, and indeed I still doubt. The subject interested me beyond what, I suppose, it is worth."

Well, my Orchis-book is a success (but I do not know whether it sells)."

In another letter to the same friend, he wrote :—

"You have pleased me much by what you say in regard to Bentham and Oliver approving of my book ; for I had got a sort of nervousness, and doubted whether I had not made an egregious fool of myself, and concocted pleasant little stinging remarks for reviews, such as 'Mr. Darwin's head seems to have been turned by a certain degree of success, and he thinks that the most trifling observations are worth publication.'"

Mr. Bentham's approval was given in his Presidential Address to the Linnean Society, May 24, 1862, and was all the more valuable, because it came from one who was by no means supposed to be favourable to Evolutionary doctrines.]

C. Darwin to Asa Gray.

Down, June 10 [1862].

MY DEAR GRAY,—Your generous sympathy makes you over-estimate what you have read of my Orchid-book. But your letter of May 18th and 26th has given me an almost foolish amount of satisfaction. The subject interested me, I knew, beyond its real value ; but I had lately got to think that I had made myself a complete fool by publishing in a semi-popular form. Now I shall confidently defy the world. I have heard that Bentham and Oliver approve of it ; but I have heard the opinion of no one else whose opinion is worth a farthing. . . . No doubt my volume contains much error : how curiously difficult it is to be accurate, though I try my utmost. Your notes have interested me beyond measure. I can now afford to d— my critics with ineffable complacency of mind. Cordial thanks for this benefit. It is surprising to me that you should have strength of mind to care for science, amidst the awful events daily occurring in your country. I daily look at the *Times* with almost as much interest as an American could do.

When will peace come? it is dreadful to think of the desolation of large parts of your magnificent country; and all the speechless misery suffered by many. I hope and think it not unlikely that we English are wrong in concluding that it will take a long time for prosperity to return to you. It is an awful subject to reflect on. . . .

[Dr. Asa Gray reviewed the book in 'Silliman's Journal,' * where he speaks, in strong terms, of the fascination which it must have for even slightly instructed readers. He made, too, some original observations on an American orchid, and these first-fruits of the subject, sent in MS. or proof sheet to my father, were welcomed by him in a letter (July 23rd):—

"Last night, after writing the above, I read the great bundle of notes. Little did I think what I had to read. What admirable observations! You have distanced me on my own hobby-horse! I have not had for weeks such a glow of pleasure as your observations gave me."

The next letter refers to the publication of the review:]

C. Darwin to Asa Gray.

Down, July 28, [1862].

MY DEAR GRAY,—I hardly know what to thank for first. Your stamps gave infinite satisfaction. I took him † first one lot, and then an hour afterwards another lot. He actually raised himself on one elbow to look at them. It was the first animation he showed. He said only: "You must thank Professor Gray awfully." In the evening after a long silence, there came out the oracular sentence: "He is awfully kind." And indeed you are, overworked as you are, to take so much trouble for our

* 'Silliman's Journal,' vol. xxiv. p. 138. Here is given an account of the fertilisation of *Platanthera Hookeri*. *P. hyperborea* is discussed in Dr. Gray's 'Enumeration' in the same volume, p. 259; also, with other species, in a second notice of the Orchid-book at p. 420.

† One of his boys who was ill.

poor dear little man.—And now I must begin the "awfullys" on my own account: what a capital notice you have published on the Orchids! It could not have been better; but I fear that you overrate it. I am very sure that I had not the least idea that you or any one would approve of it so much. I return your last note for the chance of your publishing any notice on the subject; but after all perhaps you may not think it worth while; yet in my judgment *several* of your facts, especially *Platanthera hyperborea*, are *much* too good to be merged in a review. But I have always noticed that you are prodigal in originality in your reviews. . . .

[Sir Joseph Hooker reviewed the book in the *Gardeners' Chronicle*, writing in a successful imitation of the style of Lindley, the Editor. My father wrote to Sir Joseph (Nov. 12, 1862):—

"So you did write the review in the *Gardeners' Chronicle.* Once or twice I doubted whether it was Lindley; but when I came to a little slap at R. Brown, I doubted no longer. You arch-rogue! I do not wonder you have deceived others also. Perhaps I am a conceited dog; but if so, you have much to answer for; I never received so much praise, and coming from you I value it much more than from any other."

With regard to botanical opinion generally, he wrote to Dr. Gray, "I am fairly astonished at the success of my book with botanists." Among naturalists who were not botanists, Lyell was pre-eminent in his appreciation of the book. I have no means of knowing when he read it, but in later life, as I learn from Professor Judd, he was enthusiastic in praise of the 'Fertilisation of Orchids,' which he considered "next to the 'Origin,' as the most valuable of all Darwin's works." Among the general public the author did not at first hear of many disciples, thus he wrote to his cousin Fox in September 1862: "Hardly any one not a botanist, except yourself, as far as I know, has cared for it."

A favourable notice appeared in the *Saturday Review*, October 18th, 1862; the reviewer points out that the book would escape the angry polemics aroused by the 'Origin.'* This is illustrated by a review in the *Literary Churchman*, in which only one fault is found, namely, that Mr. Darwin's expression of admiration at the contrivances in orchids is too indirect a way of saying, "O Lord, how manifold are Thy works!"

A somewhat similar criticism occurs in the 'Edinburgh Review' (October 1862). The writer points out that Mr. Darwin constantly uses phrases, such as "beautiful contrivance," "the labellum is . . . *in order* to attract," "the nectar is *purposely* lodged." The Reviewer concludes his discussion thus: "We know, too, that these purposes and ideas are not our own, but the ideas and purposes of Another."

The 'Edinburgh' reviewer's treatment of his subject was criticised in the *Saturday Review*, November 15th, 1862. With reference to this article my father wrote to Sir Joseph Hooker (December 29th, 1862):—

"Here is an odd chance; my nephew Henry Parker, an Oxford Classic, and Fellow of Oriel, came here this evening; and I asked him whether he knew who had written the little article in the *Saturday*, smashing the [Edinburgh reviewer], which we liked; and after a little hesitation he owned he had. I never knew that he wrote in the *Saturday;* and was it not an odd chance?"

The 'Edinburgh' article was written by the Duke of Argyll, and has since been made use of in his 'Reign of Law,' 1867. Mr. Wallace replied † to the Duke's criticisms, making some especially good remarks on those which refer to orchids. He shows how, by a "beautiful self-acting adjustment," the nectary of the orchid Angræcum (from 10 to 14 inches in

* Dr. Gray pointed out that if the Orchid-book (with a few trifling omissions) had appeared before the 'Origin,' the author would have been canonised rather than anathematised by the natural theologians.

† 'Quarterly Journal of Science,' October 1867. Republished in 'Natural Selection,' 1871.

length), and the proboscis of a moth sufficiently long to reach the nectar, might be developed by natural selection. He goes on to point out that on any other theory we must suppose that the flower was created with an enormously long nectary, and that then by a special act, an insect was created fitted to visit the flower, which would otherwise remain sterile. With regard to this point my father wrote (October 12 or 13, 1867):—

"I forgot to remark how capitally you turn the tables on the Duke, when you make him create the Angræcum and Moth by special creation."

If we examine the literature relating to the fertilisation of flowers, we do not find that this new branch of study showed any great activity immediately after the publication of the Orchid-book. There are a few papers by Asa Gray, in 1862 and 1863, by Hildebrand in 1864, and by Moggridge in 1865, but the great mass of work by Axell, Delpino, Hildebrand, and the Müllers, did not begin to appear until about 1867. The period during which the new views were being assimilated, and before they became thoroughly fruitful, was, however, surprisingly short. The later activity in this department may be roughly gauged by the fact that the valuable 'Bibliography,' given by Prof. D'Arcy Thompson in his translation of Müller's 'Befruchtung' (1883), contains references to 814 papers.

Besides the book on Orchids, my father wrote two or three papers on the subject, which will be found mentioned in the Appendix. The earliest of these, on the three sexual forms of Catasetum, was published in 1862; it is an anticipation of part of the Orchid-book, and was merely published in the Linnean Society's Journal, in acknowledgment of the use made of a specimen in the Society's possession. The possibility of apparently distinct species being merely sexual forms of a single species, suggested a characteristic experiment, which is alluded to in the following letter to one of his earliest disciples in the study of the fertilisation of flowers:]

*C. Darwin to J. Traherne Moggridge.**

Down, October 13 [1865].

MY DEAR SIR,—I am especially obliged to you for your beautiful plates and letter-press; for no single point in natural history interests and perplexes me so much as the self-fertilisation † of the Bee-orchis. You have already thrown some light on the subject, and your present observations promise to throw more.

I formed two conjectures: first, that some insect during certain seasons might cross the plants, but I have almost given up this; nevertheless, pray have a look at the flowers next season. Secondly, I conjectured that the Spider and Bee-orchids might be a crossing and self-fertile form of the same species. Accordingly I wrote some years ago to an acquaintance, asking him to mark some Spider-orchids, and observe whether they retained the same character; but he evidently thought the request as foolish as if I had asked him to mark one of his cows with a ribbon, to see if it would turn next spring into a horse. Now will you be so kind as to tie a string round the stem of half-a-dozen Spider-orchids, and when you leave Mentone dig them up, and I would try and cultivate them and see if they kept constant; but I should require to know in what sort of soil and situations they grow. It would be indispensable to mark the plant so that there could be no mistake about the individual. It is also just possible that the same plant would throw up, at different seasons different flower-scapes, and the marked plants would serve as evidence.

With many thanks, my dear sir,

Yours sincerely,

CH. DARWIN.

* The late Mr. Moggridge, author of 'Harvesting Ants and Trap-door Spiders,' 'Flora of Mentone,' &c.

† He once remarked to Dr. Norman Moore that one of the things that made him wish to live a few thousand years, was his desire to see the extinction of the Bee-orchis,—an end to which he believed its self-fertilising habit was leading.

P.S.—I send by this post my paper on climbing plants, parts of which you might like to read.

[Sir Thomas Farrer and Dr. W. Ogle were also guided and encouraged by my father in their observations. The following refers to a paper by Sir Thomas Farrer, in the 'Annals and Magazine of Natural History,' 1868, on the fertilisation of the Scarlet Runner:]

C. Darwin to T. H. Farrer.

Down, Sept. 15, 1868.

MY DEAR MR. FARRER,—I grieve to say that the *main* features of your case are known. I am the sinner and described them some ten years ago. But I overlooked many details, as the appendage to the single stamen, and several other points. I send my notes, but I must beg for their return, as I have *no other copy.* I quite agree, the facts are most striking, especially as you put them. Are you sure that the Hive-bee is the cutter? it is against my experience. If sure, make the point more prominent, or if not sure, erase it. I do not think the subject is quite new enough for the Linnean Society; but I dare say the 'Annals and Magazine of Natural History,' or *Gardeners' Chronicle* would gladly publish your observations, and it is a great pity they should be lost. If you like I would send your paper to either quarter with a note. In this case you must give a title, and your name, and perhaps it would be well to premise your remarks with a line of reference to my paper stating that you had observed independently and more fully.

I have read my own paper over after an interval of several years, and am amused at the caution with which I put the case that the final end was for crossing distinct individuals, of which I was then as fully convinced as now, but I knew that the doctrine would shock all botanists. Now the opinion is becoming familiar.

To see penetration of pollen-tubes is not difficult, but in most cases requires some practice with dissecting under a one-tenth of an inch focal distance single lens; and just at first this will seem to you extremely difficult.

What a capital observer you are—a first-rate Naturalist has been sacrificed, or partly sacrificed, to Public life.

Believe me, yours very sincerely,

CH. DARWIN.

P.S.—If you come across any large Salvia, look at it—the contrivance is admirable. It went to my heart to tell a man who came here a few weeks ago with splendid drawings and MS. on Salvia, that the work had been all done in Germany.*

[The following extract is from a letter, November 26th, 1868, to Sir Thomas Farrer, written as I learn from him, "in answer to a request for some advice as to the best modes of observation."

"In my opinion the best plan is to go on working and making copious notes, without much thought of publication, and then if the results turn out striking publish them. It is my impression, but I do not feel sure that I am right, that the best and most novel plan would be, instead of describing the means of fertilisation in particular plants, to investigate the part which certain structures play with all plants or throughout certain orders; for instance, the brush of hairs on the style, or the diadelphous condition of the stamens in the Leguminosæ, or the hairs within the corolla, &c. &c. Looking to your note, I think that this is perhaps the plan which you suggest.

It is well to remember that Naturalists value observations

* Dr. W. Ogle, the observer of the fertilisation of Salvia here alluded to, published his results in the 'Pop. Science Review,' 1869. He refers both gracefully and gratefully to his relationship with my father in the introduction to his translation of Kerner's 'Flowers and their Unbidden Guests.'

far more than reasoning; therefore your conclusions should be as often as possible fortified by noticing how insects actually do the work."

In 1869, Sir Thomas Farrer corresponded with my father on the fertilisation of Passiflora and of Tacsonia. He has given me his impressions of the correspondence:—

"I had suggested that the elaborate series of *chevaux-de-frise*, by which the nectary of the common Passiflora is guarded, were specially calculated to protect the flower from the stiff-beaked humming birds which would not fertilize it, and to facilitate the access of the little proboscis of the humble bee, which would do so; whilst, on the other hand, the long pendent tube and flexible valve-like corona which retains the nectar of Tacsonia would shut out the bee, which would not, and admit the humming bird which would, fertilize that flower. The suggestion is very possibly worthless, and could only be verified or refuted by examination of flowers in the countries where they grow naturally. . . . What interested me was to see that on this as on almost any other point of detailed observation, Mr. Darwin could always say, 'Yes; but at one time I made some observations myself on this particular point; and I think you will find, &c. &c.' That he should after years of interval remember that he had noticed the peculiar structure to which I was referring in the *Passiflora princeps* struck me at the time as very remarkable."

With regard to the spread of a belief in the adaptation of flowers for cross-fertilisation, my father wrote to Mr. Bentham April 22, 1868:—

"Most of the criticisms which I sometimes meet with in French works against the frequency of crossing, I am certain are the result of mere ignorance. I have never hitherto found the rule to fail that when an author describes the structure of a flower as specially adapted for self-fertilisation, it is really adapted for crossing. The Fumariaceæ offer a good instance of this, and Treviranus threw this order in my teeth; but in Corydalis, Hildebrand shows how utterly false the idea of self-fertilisation is. This author's paper on Salvia is really worth reading, and I have observed some species, and know that he is accurate."

The next letter refers to Professor Hildebrand's paper on Corydalis, published in the 'Proc. Internat. Hort. Congress,' London, 1866, and in Pringsheim's 'Jahrbücher,' vol. v. The memoir on Salvia alluded to is contained in the previous volume of the same Journal:]

*C. Darwin to F. Hildebrand.**

Down, May 16 [1866].

MY DEAR SIR,—The state of my health prevents my attending the Hort. Congress; but I forwarded yesterday your paper to the secretary, and if they are not overwhelmed with papers, yours will be gladly received. I have made many observations on the Fumariaceæ, and convinced myself that they were adapted for insect agency; but I never observed anything nearly so curious as your most interesting facts. I hope you will repeat your experiments on the Corydalis on a larger scale, and especially on several distinct plants; for your plant might have been individually peculiar, like certain individual plants of Lobelia, &c., described by Gärtner, and of Passiflora and Orchids described by Mr. Scott. . . .

Since writing to you before, I have read your admirable memoir on Salvia, and it has interested me almost as much as when I first investigated the structure of Orchids. Your paper illustrates several points in my 'Origin of Species,' especially the transition of organs. Knowing only two or three species in the genus, I had often marvelled how one cell of the anther could have been transformed into the movable plate or spoon; and how well you show the gradations;

* Professor of Botany at Freiburg.

but I am surprised that you did not more strongly insist on this point.

I shall be still more surprised if you do not ultimately come to the same belief with me, as shown by so many beautiful contrivances, that all plants require, from some unknown cause, to be occasionally fertilized by pollen from a distinct individual. With sincere respect, believe me, my dear Sir,

Yours very faithfully,

CH. DARWIN.

[The following letter refers to the late Hermann Müller's 'Befruchtung der Blumen,' by far the most valuable of the mass of literature originating in the 'Fertilisation of Orchids.' An English translation, by Prof. D'Arcy Thompson was published in 1883. My father's "Prefatory Notice" to this work is dated February 6, 1882, and is therefore almost the last of his writings:]

C. Darwin to H. Müller.

Down, May 5, 1873.

MY DEAR SIR,—Owing to all sorts of interruptions and to my reading German so slowly, I have read only to p. 88 of your book; but I must have the pleasure of telling you how very valuable a work it appears to me. Independently of the many original observations, which of course form the most important part, the work will be of the highest use as a means of reference to all that has been done on the subject. I am fairly astonished at the number of species of insects, the visits of which to different flowers you have recorded. You must have worked in the most indefatigable manner. About half a year ago the editor of 'Nature' suggested that it would be a grand undertaking if a number of naturalists were to do what you have already done on so large a scale with respect to the visits of insects. I have been particularly glad to read your historical sketch, for I had never before seen all the references put together. I have sometimes feared that I was in error when I said that C. K. Sprengel did not fully perceive that cross-fertilisation was the final end of the structure of flowers; but now this fear is relieved, and it is a great satisfaction to me to believe that I have aided in making his excellent book more generally known. Nothing has surprised me more than to see in your historical sketch how much I myself have done on the subject, as it never before occurred to me to think of all my papers as a whole. But I do not doubt that your generous appreciation of the labours of others has led you to over-estimate what I have done. With very sincere thanks and respect, believe me,

Yours faithfully,

CHARLES DARWIN.

P.S.—I have mentioned your book to almost every one who, as far as I know, cares for the subject in England; and I have ordered a copy to be sent to our Royal Society.

[The next letter, to Dr. Behrens, refers to the same subject as the last:]

C. Darwin to W. Behrens.

Down, August 29 [1878].

DEAR SIR,—I am very much obliged to you for having sent me your 'Geschichte der Bestaubungs-Theorie,' * and which has interested me much. It has put some things in a new light, and has told me other things which I did not know. I heartily agree with you in your high appreciation of poor old C. Sprengel's work; and one regrets bitterly that he did not live to see his labours thus valued. It rejoices me also to notice how highly you appreciate H. Müller, who has always seemed to me an admirable observer and reasoner. I am at present endeavouring to persuade an English publisher to bring out a translation of his 'Befruchtung.'

* Progr. der K. Gewerbschule zu Elberfeld, 1877, 1878.

Lastly, permit me to thank you for your very generous remarks on my works. By placing what I have been able to do on this subject in systematic order, you have made me think more highly of my own work than I ever did before! Nevertheless, I fear that you have done me more than justice.

I remain, dear Sir, yours faithfully and obliged,

CHARLES DARWIN.

[The letter which follows was called forth by Dr. Gray's article in 'Nature,' to which reference has already been made, and which appeared June 4, 1874:]

C. Darwin to Asa Gray.

Down, June 3 [1874].

MY DEAR GRAY,—I was rejoiced to see your handwriting again in your note of the 4th, of which more anon. I was astonished to see announced about a week ago that you were going to write in 'Nature' an article on me, and this morning I received an advance copy. It is the grandest thing ever written about me, especially as coming from a man like yourself. It has deeply pleased me, particularly some of your side remarks. It is a wonderful thing to me to live to see my name coupled in any fashion with that of Robert Brown. But you are a bold man, for I am sure that you will be sneered at by not a few botanists. I have never been so honoured before, and I hope it will do me good and make me try to be as careful as possible; and good heavens, how difficult accuracy is! I feel a very proud man, but I hope this won't last. . . .

[Fritz Müller has observed that the flowers of Hedychium are so arranged that the pollen is removed by the wings of hovering butterflies. My father's prediction of this observation is given in the following letter:—]

C. Darwin to H. Müller.

Down, August 7, 1876.

. . . . I was much interested by your brother's article on Hedychium; about two years ago I was so convinced that the flowers were fertilized by the tips of the wings of large moths, that I wrote to India to ask a man to observe the flowers and catch the moths at work, and he sent me 20 to 30 Sphinx-moths, but so badly packed that they all arrived in fragments; and I could make out nothing. . . .

Yours sincerely,
CH. DARWIN.

[The following extract from a letter (Feb. 25, 1864), to Dr. Gray refers to another prediction fulfilled:—

"I have of course seen no one, and except good dear Hooker, I hear from no one. He, like a good and true friend, though so overworked, often writes to me.

"I have had one letter which has interested me greatly, with a paper, which will appear in the Linnean Journal, by Dr. Crüger of Trinidad, which shows that I am all right about Catasetum, even to the spot where the pollinia adhere to the bees, which visit the flower, as I said, to gnaw the labellum. Crüger's account of Coryanthes and the use of the bucket-like labellum full of water beats everything: I *suspect* that the bees being well wetted flattens their hairs, and allows the viscid disc to adhere."]

C. Darwin to the Marquis de Saporta.

Down, December 24, 1877.

MY DEAR SIR,—I thank you sincerely for your long and most interesting letter, which I should have answered sooner had it not been delayed in London. I had not heard before that I was to be proposed as a Corresponding Member of the Institute. Living so retired a life as I do, such honours affect me very little, and I can say with entire truth that your kind expression of sympathy has given and will give me much more pleasure than the election itself, should I be elected.

Your idea that dicotyledonous plants were not developed in force until sucking insects had been evolved seems to me a splendid one. I am surprised that the idea never occurred to me, but this is always the case when one first hears a new and simple explanation of some mysterious phenomenon . . . . I formerly showed that we might fairly assume that the beauty of flowers, their sweet odour and copious nectar, may be attributed to the existence of flower-haunting insects, but your idea, which I hope you will publish, goes much further and is much more important. With respect to the great development of mammifers in the later Geological periods following from the development of dicotyledons, I think it ought to be proved that such animals as deer, cows, horses, &c. could not flourish if fed exclusively on the gramineæ and other anemophilous monocotyledons ; and I do not suppose that any evidence on this head exists.

Your suggestion of studying the manner of fertilisation of the surviving members of the most ancient forms of the dicotyledons is a very good one, and I hope that you will keep it in mind yourself, for I have turned my attention to other subjects. Delpino I think says that Magnolia is fertilised by insects which gnaw the petals, and I should not be surprised if the same fact holds good with Nymphæa. Whenever I have looked at the flowers of these latter plants I have felt inclined to admit the view that petals are modified stamens, and not modified leaves ; though Poinsettia seems to show that true leaves might be converted into coloured petals. I grieve to say that I have never been properly grounded in Botany and have studied only special points—therefore I cannot pretend to express any opinion on your remarks on the origin of the flowers of the Coniferæ, Gnetaceæ, &c; but I have been delighted with what you say on the conversion of a monœcious species into a hermaphrodite one by the condensations of the verticils on a branch bearing female flowers near the summit, and male flowers below.

I expect Hooker to come here before long, and I will then show him your drawing, and if he makes any important remarks I will communicate with you. He is very busy at present in clearing off arrears after his American Expedition, so that I do not like to trouble him, even with the briefest note. I am at present working with my son at some Physiological subjects, and we are arriving at very curious results, but they are not as yet sufficiently certain to be worth communicating to you. . . .

[In 1877 a second edition of the 'Fertilisation of Orchids' was published, the first edition having been for some time out of print. The new edition was remodelled and almost rewritten, and a large amount of new matter added, much of which the author owed to his friend Fritz Müller.

With regard to this edition he wrote to Dr. Gray:—

"I do not suppose I shall ever again touch the book. After much doubt I have resolved to act in this way with all my books for the future; that is to correct them once and never touch them again, so as to use the small quantity of work left in me for new matter."

He may have felt a diminution of his power of reviewing large bodies of facts, such as would be needed in the preparation of new editions, but his powers of observation were certainly not diminished. He wrote to Mr. Dyer on July 14, 1878:—]

MY DEAR DYER,—*Thalia dealbata* was sent me from Kew: it has flowered and after looking casually at the flowers, they have driven me almost mad, and I have worked at them for a week: it is as grand a case as that of Catasetum.

Pistil vigorously motile (so that whole flower shakes when pistil suddenly coils up); when excited by a touch the two filaments [are] produced laterally and transversely across the flower (just over the nectar) from one of the petals or modified stamens. It is splendid to watch the phenomenon under a weak power when a bristle is inserted into a *young* flower which no insect has visited. As far as I know Stylidium is the sole case of sensitive pistil and here it is the pistil + stamens. In Thalia * cross-fertilisation is ensured by the wonderful movement, if bees visit several flowers.

I have now relieved my mind and will tell the purport of this note—viz. if any other species of Thalia besides *T. dealbata* should flower with you, for the love of heaven and all the saints, send me a few in *tin box with damp moss.*

Your insane friend,
CH. DARWIN.

[In 1878 Dr. Ogle's translation of Kerner's interesting book, 'Flowers and their Unbidden Guests,' was published. My father, who felt much interest in the translation (as appears in the following letter), contributed some prefatory words of approval :]

C. Darwin to W. Ogle.

Down, December 16 [1878].

. . . . I have now read Kerner's book, which is better even than I anticipated. The translation seems to me as clear as daylight, and written in forcible and good familiar English. I am rather afraid that it is too good for the English public, which seems to like very washy food, unless it be administered by some one whose name is well known, and then I suspect a good deal of the unintelligible is very pleasing to them. I hope to heaven that I may be wrong.

* Hildebrand has described an explosive arrangement in some of the Maranteæ—the tribe to which Thalia belongs.

Anyhow, you and Mrs. Ogle have done a right good service for Botanical Science.

Yours very sincerely,
CH. DARWIN.

P.S.—You have done me much honour in your prefatory remarks.

[One of the latest references to his Orchid-work occurs in a letter to Mr. Bentham, February 16, 1880. It shows the amount of pleasure which this subject gave to my father, and (what is characteristic of him) that his reminiscence of the work was one of delight in the observations which preceded its publication, not to the applause which followed it :—

" They are wonderful creatures, these Orchids, and I sometimes think with a glow of pleasure, when I remember making out some little point in their method of fertilisation."]

CHAPTER VIII.

THE 'EFFECTS OF CROSS- AND SELF-FERTILISATION IN THE VEGETABLE KINGDOM.' 1876.

[THIS book, as pointed out in the 'Autobiography,' is a complement to the 'Fertilisation of Orchids,' because it shows how important are the results of cross-fertilisation which are ensured by the mechanisms described in that book. By proving that the offspring of cross-fertilisation are more vigorous than the offspring of self-fertilisation, he showed that one circumstance which influences the fate of young plants in the struggle for life is the degree to which their parents are fitted for cross-fertilisation. He thus convinced himself that the intensity of the struggle (which he had elsewhere shown to exist among young plants) is a measure of the strength of a selective agency perpetually sifting out every modification in the structure of flowers which can affect its capabilities for cross-fertilisation.

The book is also valuable in another respect, because it throws light on the difficult problems of the origin of sexuality. The increased vigour resulting from cross-fertilisation is allied in the closest manner to the advantage gained by change of conditions. So strongly is this the case, that in some instances cross-fertilisation gives no advantage to the offspring, unless the parents have lived under slightly different conditions. So that the really important thing is not that two individuals of different *blood* shall unite, but two individuals which have been subjected to different conditions. We are thus led to believe that sexuality is a means for infusing vigour into the offspring by the coalescence of differentiated elements, an advantage which could not follow if reproductions were entirely asexual.

It is remarkable that this book, the result of eleven years of experimental work, owed its origin to a chance observation. My father had raised two beds of *Linaria vulgaris*—one set being the offspring of cross- and the other of self-fertilisation. These plants were grown for the sake of some observations on inheritance, and not with any view to cross-breeding, and he was astonished to observe that the offspring of self-fertilisation were clearly less vigorous than the others. It seemed incredible to him that this result could be due to a single act of self-fertilisation, and it was only in the following year, when precisely the same result occurred in the case of a similar experiment on inheritance in Carnations, that his attention was "thoroughly aroused," and that he determined to make a series of experiments specially directed to the question. The following letters give some account of the work in question:]

C. Darwin to Asa Gray.

September 10, [1866?]

. . . . I have just begun a large course of experiments on the germination of the seed, and on the growth of the young plants when raised from a pistil fertilised by pollen from the same flower, and from pollen from a distinct plant of the same, or of some other variety. I have not made sufficient experiments to judge certainly, but in some cases the difference in the growth of the young plants is highly remarkable. I have taken every kind of precaution in getting seed from the same plant, in germinating the seed on my own chimney-piece, in planting the seedlings in the same flower-pot, and under this similar treatment I have seen the young seedlings from the crossed seed exactly twice as tall as the seedlings from the self-fertilised seed; both seeds having germinated on same day. If I can establish this fact (but perhaps it will all go to the dogs), in some fifty cases, with plants of different orders, I think it will be very important, for then we shall positively know why the structure of every flower permits, or favours, or necessitates an occasional cross with a distinct individual. But all this is rather cooking my hare before I have caught it. But somehow it is a great pleasure to me to tell you what I am about.

Believe me, my dear Gray,
Ever yours most truly, and with cordial thanks,
CH. DARWIN.

C. Darwin to G. Bentham.

April 22, 1868.

. . . . I am experimenting on a very large scale on the difference in power of growth between plants raised from self-fertilised and crossed seeds; and it is no exaggeration to say that the difference in growth and vigour is sometimes truly wonderful. Lyell, Huxley and Hooker have seen some of my plants, and been astonished; and I should much like to show them to you. I always supposed until lately that no evil effects would be visible until after several generations of self-fertilisation; but now I see that one generation sometimes suffices; and the existence of dimorphic plants and all the wonderful contrivances of orchids are quite intelligible to me.

With cordial thanks for your letter, which has pleased me greatly,

Yours very sincerely,
CHARLES DARWIN.

[An extract from a letter to Dr. Gray (March 11, 1873) mentions the progress of the work:—

"I worked last summer hard at Drosera, but could not finish till I got fresh plants, and consequently took up the effects of crossing and self-fertilising plants, and am got so interested that Drosera must go to the dogs till I finish with this, and get it published; but then I will resume my beloved Drosera, and I heartily apologise for having sent the precious little things even for a moment to the dogs."

The following letters give the author's impression of his own book.]

C. Darwin to J. Murray.

Down, September 16, 1876.

MY DEAR SIR,—I have just received proofs in sheet of five sheets, so you will have to decide soon how many copies will have to be struck off. I do not know what to advise. The greater part of the book is extremely dry, and the whole on a special subject. Nevertheless, I am convinced that the book is of value, and I am convinced that for *many* years copies will be occasionally sold. Judging from the sale of my former books, and from supposing that some persons will purchase it to complete the set of my works, I would suggest 1500. But you must be guided by your larger experience. I will only repeat that I am convinced the book is of some permanent value. . . .

C. Darwin to Victor Carus.

Down, September 27, 1876.

MY DEAR SIR,—I sent by this morning's post the four first perfect sheets of my new book, the title of which you will see on the first page, and which will be published early in November.

I am sorry to say that it is only shorter by a few pages than my 'Insectivorous Plants.' The whole is now in type, though I have corrected finally only half the volume. You will, therefore, rapidly receive the remainder. The book is very dull. Chapters II. to VI., inclusive, are simply a record of experiments. Nevertheless, I believe (though a man can never judge his own books) that the book is valuable. You will have to decide whether it is worth translating. I hope so. It has cost me very great labour, and the results seem to me remarkable and well established.

If you translate it, you could easily get aid for Chapters II. to VI., as there is here endless, but, I have thought, necessary repetition. I shall be anxious to hear what you decide. . . . . .

I most sincerely hope that your health has been fairly good this summer.

My dear Sir, yours very truly,

CH. DARWIN.

C. Darwin to Asa Gray.

Down, October 28, 1876.

MY DEAR GRAY,—I send by this post all the clean sheets as yet printed, and I hope to send the remainder within a fortnight. Please observe that the first six chapters are not readable, and the six last very dull. Still I believe that the results are valuable. If you review the book, I shall be very curious to see what you think of it, for I care more for your judgment than for that of almost any one else. I know also that you will speak the truth, whether you approve or disapprove. Very few will take the trouble to read the book, and I do not expect you to read the whole, but I hope you will read the latter chapters.

. . . I am so sick of correcting the press and licking my horrid bad style into intelligible English.

[The 'Effects of Cross and Self-Fertilisation' was published on November 10, 1876, and 1500 copies were sold before the end of the year. The following letter refers to a review in 'Nature:' *]

* February 15, 1877.

C. Darwin to W. Thiselton Dyer.

Down, February 16, 1877.

DEAR DYER,—I must tell you how greatly I am pleased and honoured by your article in 'Nature,' which I have just read. You are an adept in saying what will please an author, not that I suppose you wrote with this express intention. I should be very well contented to deserve a fraction of your praise. I have also been much interested, and this is better than mere pleasure, by your argument about the separation of the sexes. I dare say that I am wrong, and will hereafter consider what you say more carefully: but at present I cannot drive out of my head that the sexes must have originated from two individuals, slightly different, which conjugated. But I am aware that some cases of conjugation are opposed to any such views.

With hearty thanks,

Yours sincerely,

CHARLES DARWIN.

CHAPTER IX.

'DIFFERENT FORMS OF FLOWERS ON PLANTS OF THE SAME SPECIES.' 1877.

[THE volume bearing the above title was published in 1877, and was dedicated by the author to Professor Asa Gray, "as a small tribute of respect and affection." It consists of certain earlier papers re-edited, with the addition of a quantity of new matter. The subjects treated in the book are:—

(i.) Heterostyled Plants.

(ii.) Polygamous, Diœcious, and Gynodiœcious Plants.

(iii.) Cleistogamic Flowers.

The nature of heterostyled plants may be illustrated in the primrose, one of the best known examples of the class. If a number of primroses be gathered, it will be found that some plants yield nothing but "pin-eyed" flowers, in which the style (or organ for the transmission of the pollen to the ovule) is long, while the others yield only "thrum-eyed" flowers with short styles. Thus primroses are divided into two sets or castes differing structurally from each other. My father showed that they also differ sexually, and that in fact the bond between the two castes more nearly resembles that between separate sexes than any other known relationship. Thus for example a long-styled primrose, though it can be fertilised by its own pollen, is not *fully* fertile unless it is impregnated by the pollen of a short-styled flower. Heterostyled plants are comparable to hermaphrodite animals, such as snails, which require the concourse of two individuals, although each possesses both the sexual elements. The difference is that in the case of the primrose it is *perfect fertility*, and not simply *fertility*, that depends on the mutual action of the two sets of individuals.

The work on heterostyled plants has a special bearing, to which the author attached much importance, on the problem of origin of species.*

He found that a wonderfully close parallelism exists between hybridisation and certain forms of fertilisation among heterostyled plants. So that it is hardly an exaggeration to say that the "illegitimately" reared seedlings are hybrids, although both their parents belong to identically the same species. In a letter to Professor Huxley, given in the second volume (p. 384), my father writes as if his researches on heterostyled plants tended to make him believe that sterility is a selected or acquired quality. But in his later publications, *e.g.* in the sixth edition of the 'Origin,' he adheres to the belief that sterility is an incidental rather than a selected quality. The result of his work on heterostyled plants is of importance as showing that sterility is no test of specific distinctness, and that it depends on differentiation of the sexual elements which is independent of any racial difference. I imagine that it was his instinctive love of making out a difficulty which to a great extent kept him at work so patiently on the heterostyled plants. But it was the fact that general conclusions of the above character could be drawn from his results which made him think his results worthy of publication.†

The papers which on this subject preceded and contributed to 'Forms of Flowers' were the following :—

"On the two Forms or Dimorphic Condition in the Species of Primula, and on their remarkable Sexual Relations." Linn. Soc. Journal, 1862.

* See 'Autobiography,' vol. i. p. 97.

† See 'Forms of Flowers,' p. 243.

"On the Existence of Two Forms, and on their Reciprocal Sexual Relations, in several Species of the Genus Linum." Linn. Soc. Journal, 1863.

"On the Sexual Relations of the Three Forms of *Lythrum salicaria*," Ibid. 1864.

"On the Character and Hybrid-like Nature of the Offspring from the Illegitimate Unions of Dimorphic and Trimorphic Plants." Ibid. 1869.

On the Specific Differences between *Primula veris*, Brit. Fl. (var *officinalis*, Linn.), *P. vulgaris*, Brit. Fl. (var. *acaulis*, Linn.), and *P. elatior*, Jacq.; and on the Hybrid Nature of the Common Oxlip. With Supplementary Remarks on Naturally Produced Hybrids in the Genus Verbascum." Ibid. 1869.

The following letter shows that he began the work on heterostyled plants with an erroneous view as to the meaning of the facts.]

C. Darwin to J. D. Hooker.

Down, May 7 [1860].

. . . . I have this morning been looking at my experimental cowslips, and I find some plants have all flowers with long stamens and short pistils, which I will call "male plants," others with short stamens and long pistils, which I will call "female plants." This I have somewhere seen noticed, I think by Henslow; but I find (after looking at my two sets of plants) that the stigmas of the male and female are of slightly different shape, and certainly different degree of roughness, and what has astonished me, the pollen of the so-called female plant, though very abundant, is more transparent, and each granule is exactly only $\frac{2}{3}$ of the size of the pollen of the so-called male plants. Has this been observed? I cannot help suspecting [that] the cowslip is in fact diœcious, but it may turn out all a blunder, but anyhow I will mark with sticks the so-called male and female plants and watch their seeding. It would be a fine case of gradation between an hermaphrodite and unisexual condition. Likewise a sort of case of balancement of long and short pistils and stamens. Likewise perhaps throws light on oxlips. . . .

I have now examined primroses and find exactly the same difference in the size of the pollen, correlated with the same difference in the length of the style and roughness of the stigmas.

C. Darwin to Asa Gray.

June 8 [1860].

. . . . I have been making some little trifling observations which have interested and perplexed me much. I find with primroses and cowslips, that about an equal number of plants are thus characterised.

So-called (by me) *male* plant. Pistil much shorter than stamens; stigma rather smooth,—*pollen grains large*, throat of corolla short.

So-called female plant. Pistil much longer than stamens, stigma rougher, *pollen-grains smaller*,—throat of corolla long.

I have marked a lot of plants, and expected to find the so-called male plant barren; but judging from the feel of the capsules, this is not the case, and I am very much surprised at the difference in the size of the pollen. . . . If it should prove that the so-called male plants produce less seed than the so-called females, what a beautiful case of gradation from hermaphrodite to unisexual condition it will be! If they produce about equal number of seed, how perplexing it will be.

C. Darwin to J. D. Hooker.

Down, December 17, [1860?]

. . . . I have just been ordering a photograph of myself for a friend; and have ordered one for you, and for heaven's sake oblige me, and burn that now hanging up in your room.—It makes me look atrociously wicked.

. . . . In the spring I must get you to look for long pistils and short pistils in the rarer species of Primula and in some allied Genera. It holds with *P. Sinensis*. You remember all the fuss I made on this subject last spring; well, the other day at last I had time to weigh the seeds, and by Jove the plants of primrose and cowslip with short pistils and large grained pollen * are rather more fertile than those with long pistils, and small-grained pollen. I find that they require the action of insects to set them, and I never will believe that these differences are without some meaning.

Some of my experiments lead me to suspect that the large-grained pollen suits the long pistils and the small-grained pollen suits the short pistils; but I am determined to see if I cannot make out the mystery next spring.

How does your book on plants brew in your mind? Have you begun it? . . .

Remember me most kindly to Oliver. He must be astonished at not having a string of questions, I fear he will get out of practice!

[The Primula-work was finished in the autumn of 1861, and on Nov. 8th he wrote to Sir J. D. Hooker:—

"I have sent my paper on dimorphism in Primula to the Linn. Soc. I shall go up and read it whenever it comes on; I hope you may be able to attend, for I do not suppose many will care a penny for the subject."

With regard to the reading of the paper (on Nov. 21st), he wrote to the same friend:—

"I by no means thought that I produced a "tremendous effect" in the Linn. Soc., but by Jove the Linn. Soc., produced a tremendous effect on me, for I could not get out of bed till late next evening, so that I just crawled home. I fear I must give up trying to read any paper or speak; it is a horrid bore, I can do nothing like other people.

* Thus the plants which he imagined to be tending towards a male condition were more productive than the supposed females.

To Dr. Gray he wrote, (Dec. 1861):—

"You may rely on it, I will send you a copy of my Primula paper as soon as I can get one; but I believe it will not be printed till April 1st, and therefore after my Orchid Book. I care more for your and Hooker's opinion than for that of all the rest of the world, and for Lyell's on geological points. Bentham and Hooker thought well of my paper when read; but no one can judge of evidence by merely hearing a paper."

The work on Primula was the means of bringing my father in contact with the late Mr. John Scott, then working as a gardener in the Botanic Gardens at Edinburgh,—an employment which he seems to have chosen in order to gratify his passion for natural history. He wrote one or two excellent botanical papers, and ultimately obtained a post in India.* He died in 1880.

A few phrases may be quoted from letters to Sir J. D. Hooker, showing my father's estimate of Scott:—

"If you know, do please tell me who is John Scott of the Botanical Gardens of Edinburgh; I have been corresponding largely with him; he is no common man."

"If he had leisure he would make a wonderful observer; to my judgment I have come across no one like him."

"He has interested me strangely, and I have formed a very high opinion of his intellect. I hope he will accept pecuniary assistance from me; but he has hitherto refused." (He ultimately succeeded in being allowed to pay for Mr. Scott's passage to India.)

"I know nothing of him excepting from his letters; these show remarkable talent, astonishing perseverance, much modesty, and what I admire, determined difference from me on many points."

So highly did he estimate Scott's abilities that he formed

* While in India he made some admirable observations on expression for my father.

a plan (which however never went beyond an early stage of discussion) of employing him to work out certain problems connected with intercrossing.

The following letter refers to my father's investigations on Lythrum,* a plant which reveals even a more wonderful condition of sexual complexity than that of Primula. For in Lythrum there are not merely two, but three castes, differing structurally and physiologically from each other :]

C. Darwin to Asa Gray.

Down, August 9 [1862].

My dear Gray,—It is late at night, and I am going to write briefly, and of course to beg a favour.

The Mitchella very good, but pollen apparently equal-sized. I have just examined Hottonia, grand difference in pollen. *Echium vulgare*, a humbug, merely a case like Thymus. But I am almost stark staring mad over Lythrum ;† if I can prove what I fully believe; it is a grand case of Trimorphism, with three different pollens and three stigmas; I have castrated and fertilised above ninety flowers, trying all the eighteen distinct crosses which are possible within the limits of this one species! I cannot explain, but I feel sure you would think it a grand case. I have been writing to Botanists to see if I can possibly get *L. hyssopifolia*, and it has just flashed on me that you might have Lythrum in North America, and I have looked to your Manual. For the love

* He was led to this, his first case of trimorphism, by Lecoq's 'Géographie Botanique,' and this must have consoled him for the trick this work played him in turning out to be so much larger than he expected. He wrote to Sir J. D. Hooker: "Here is a good joke: I saw an extract from Lecoq, 'Géograph. Bot.,' and ordered it and hoped that it was a good sized pamphlet, and nine thick volumes have arrived!"

† On another occasion he wrote (to Dr. Gray) with regard to Lythrum: "I must hold hard, otherwise I shall spend my life over dimorphism."

of heaven have a look at some of your species, and if you can get me seed, do; I want much to try species with few stamens, if they are dimorphic; *Nesæa verticillata* I should expect to be trimorphic. Seed! Seed! Seed! I should rather like seed of Mitchella. But oh, Lythrum!

Your utterly mad friend,

C. DARWIN.

P.S.—There is reason in my madness, for I can see that to those who already believe in change of species, these facts will modify to a certain extent the whole view of Hybridity.*

[On the same subject he wrote to Sir Joseph Hooker in August 1862:—

"Is Oliver at Kew? When I am established at Bournemouth I am completely mad to examine any fresh flowers of any Lythraceous plant, and I would write and ask him if any are in bloom."

Again he wrote to the same friend in October:—

"If you ask Oliver, I think he will tell you I have got a real odd case in Lythrum, it interests me extremely, and seems to me the strangest case of propagation recorded amongst plants or animals, viz. a necessary triple alliance between three hermaphrodites. I feel sure I can now prove the truth of the case from a multitude of crosses made this summer."

* A letter to Dr. Gray (July, 1862) bears on this point: "A few days ago I made an observation which has surprised me more than it ought to do—it will have to be repeated several times, but I have scarcely a doubt of its accuracy. I stated in my Primula paper that the long-styled form of *Linum grandiflorum* was utterly sterile with its own pollen; I have lately been putting the pollen of the two forms on the division of the stigma of the *same* flower; and it strikes me as truly wonderful, that the stigma distinguishes the pollen; and is penetrated by the tubes of the one and not by those of the other; nor are the tubes exserted. Or (which is the same thing) the stigma of the one form acts on and is acted on by pollen, which produces not the least effect on the stigma of the other form. Taking sexual power as the criterion of difference, the two forms of this one species may be said to be generically distinct."

In an article, 'Dimorphism in the Genitalia of Plants' ('Silliman's Journal,' 1862, vol. xxxiv. p. 419), Dr. Gray points out that the structural difference between the two forms of Primula had already been defined in the 'Flora of N. America,' as *diœcio-dimorphism.* The use of this term called forth the following remarks from my father. The letter also alludes to a review of the 'Fertilisation of Orchids' in the same volume of 'Silliman's Journal.']

C. Darwin to Asa Gray.

Down, November 26 [1862].

My dear Gray,—The very day after my last letter, yours of November 10th, and the review in 'Silliman,' which I feared might have been lost, reached me. We were all very much interested by the political part of your letter; and in some odd way one never feels that information and opinions printed in a newspaper come from a living source; they seem dead, whereas all that you write is full of life. The reviews interested me profoundly; you rashly ask for my opinion, and you must consequently endure a long letter. First for Dimorphism; I do not *at present* like the term "Diœcio-dimorphism;" for I think it gives quite a false notion, that the phenomena are connected with a separation of the sexes. Certainly in Primula there is unequal fertility in the two forms, and I suspect this is the case with Linum; and, therefore, I felt bound in the Primula paper to state that it might be a step towards a diœcious condition; though I believe there are no diœcious forms in Primulaceæ or Linaceæ. But the three forms in Lythrum convince me that the phenomenon is in no way necessarily connected with any tendency to separation of sexes. The case seems to me in result or function to be almost identical with what old C. K. Sprengel called "dichogamy," and which is so frequent in truly hermaphrodite groups; namely, the pollen and stigma of each flower being mature at different periods. If I am right, it is very advisable not to use the term "diœcious," as this at once brings notions of separation of sexes.

. . . I was much perplexed by Oliver's remarks in the 'Natural History Review' on the Primula case, on the lower plants having sexes more often of the separated than in the higher plants,—so exactly the reverse of what takes place in animals. Hooker in his review of the 'Orchids' repeats this remark. There seems to be much truth in what you say,* and it did not occur to me, about no improbability of specialisation in *certain* lines in lowly organised beings. I could hardly doubt that the hermaphrodite state is the aboriginal one. But how is it in the conjugation of Confervæ—is not one of the two individuals here in fact male, and the other female? I have been much puzzled by this contrast in sexual arrangements between plants and animals. Can there be anything in the following consideration: By *roughest* calculation about one-third of the British *genera* of aquatic plants belong to the Linnean classes of Mono and Diœcia; whilst of terrestrial plants (the aquatic genera being subtracted) only one-thirteenth of the genera belong to these two classes. Is there any truth in this fact generally? Can aquatic plants, being confined to a small area or small community of individuals, require more free crossing, and therefore have separate sexes? But to return to one point, does not Alph. de Candolle say that aquatic plants taken as a whole are lowly organised, compared with terrestrial; and may not Oliver's remark on the separation of the sexes in lowly organised plants stand in some relation to their being frequently aquatic? Or is this all rubbish?

. . . . What a magnificent compliment you end your review with! You and Hooker seem determined to turn my head

* "Forms which are low in the scale as respects morphological completeness may be high in the scale of rank founded on specialisation of structure and function."—Dr. Gray, in 'Silliman's Journal.'

with conceit and vanity (if not already turned), and make me an unbearable wretch.

With most cordial thanks, my good and kind friend,

Farewell,

C. DARWIN.

[The following passage from a letter (July 28, 1863), to Prof. Hildebrand, contains a reference to the reception of the dimorphic work in France:—

" I am extremely much pleased to hear that you have been looking at the manner of fertilisation of your native Orchids, and still more pleased to hear that you have been experimenting on Linum. I much hope that you may publish the result of these experiments; because I was told that the most eminent French botanists of Paris said that my paper on Primula was the work of imagination, and that the case was so improbable they did not believe in my results."]

C. Darwin to Asa Gray.

April 19 [1864].

. . . . I received a little time ago a paper with a good account of your Herbarium and Library, and a long time previously your excellent review of Scott's 'Primulaceæ,' and I forwarded it to him in India, as it would much please him. I was very glad to see in it a new case of Dimorphism (I forget just now the name of the plant); I shall be grateful to hear of any other cases, as I still feel an interest in the subject. I should be very glad to get some seed of your dimorphic Plantagos; for I cannot banish the suspicion that they must belong to a very different class like that of the common Thyme.* How could the wind, which is the agent of fertilisation, with Plantago, fertilise "reciprocally dimorphic" flowers like Primula? Theory says this cannot be, and in such cases

* In this prediction he was right. See 'Forms of Flowers,' p. 307.

of one's own theories I follow Agassiz and declare, " that nature never lies." I should even be very glad to examine the two dried forms of Plantago. Indeed, any dried dimorphic plants would be gratefully received. . . .

Did my Lythrum paper interest you? I crawl on at the rate of two hours per diem, with 'Variation under Domestication.'

C. Darwin to J. D. Hooker.

Down, November 26 [1864].

. . . . You do not know how pleased I am that you have read my Lythrum paper; I thought you would not have time, and I have for long years looked at you as my Public, and care more for your opinion than that of all the rest of the world. I have done nothing which has interested me so much as Lythrum, since making out the complemental males of Cirripedes. I fear that I have dragged in too much miscellaneous matter into the paper.

. . . I get letters occasionally, which show me that Natural Selection is making *great* progress in Germany, and some amongst the young in France. I have just received a pamphlet from Germany, with the complimentary title of " Darwinische Arten-Enstehung-Humbug "!

Farewell, my best of old friends,

C. Darwin.

C. Darwin to Asa Gray.

September 10, [1867?]

. . . . The only point which I have made out this summer, which could possibly interest you, is that the common Oxlip found everywhere, more or less commonly in England, is certainly a hybrid between the primrose and cowslip; whilst the *P. elatior* (Jacq.), found only in the Eastern Counties, is a perfectly distinct and good species; hardly distinguishable from the common oxlip, except by the length of the seed-capsule relatively to the calyx. This seems to me rather a horrid fact for all systematic botanists. . . . .

C. Darwin to F. Hildebrand.

Down, November 16, 1868.

MY DEAR SIR,—I wrote my last note in such a hurry from London, that I quite forgot what I chiefly wished to say, namely to thank you for your excellent notices in the 'Bot. Zeitung' of my paper on the offspring of dimorphic plants. The subject is so obscure that I did not expect that any one would have noticed my paper, and I am accordingly very much pleased that you should have brought the subject before the many excellent naturalists of Germany.

Of all the German authors (but they are not many) whose works I have read, you write by far the clearest style, but whether this is a compliment to a German writer I do not know.

[The two following letters refer to the small bud-like "Cleistogamic" flowers found in the violet and many other plants. They do not open and are necessarily self-fertilised :]

C. Darwin to J. D. Hooker.

Down, May 30 [1862].

. . . . What will become of my book on Variation? I am involved in a multiplicity of experiments. I have been amusing myself by looking at the small flowers of Viola. If Oliver * has had time to study them, he will have seen the curious case (as it seems to me) which I have just made clearly out, viz. that in these flowers, the *few* pollen grains are

* Shortly afterwards he wrote: "Oliver, the omniscient, has sent me a paper in the 'Bot. Zeitung,' with most accurate description of all that I saw in Viola."

never shed, or never leave the anther-cells, but emit long pollen tubes, which penetrate the stigma. To-day I got the anther with the included pollen grain (now empty) at one end, and a bundle of tubes penetrating the stigmatic tissue at the other end; I got the whole under a microscope without breaking the tubes; I wonder whether the stigma pours some fluid into the anther so as to excite the included grains. It is a rather odd case of correlation, that in the double sweet violet the little flowers arc double; *i.e.*, have a multitude of minute scales representing the petals. What queer little flowers they are.

Have you had time to read poor dear Henslow's life? it has interested me for the man's sake, and, what I did not think possible, has even exalted his character in my estimation. . . . .

[The following is an extract from the letter given in part at p. 303, and refers to Dr. Gray's article on the sexual differences of plants :]

C. Darwin to Asa Gray.

November 26 [1862].

. . . . You will think that I am in the most unpleasant, contradictory, fractious humour, when I tell you that I do not like your term of "precocious fertilisation" for your second class of dimorphism [*i.e.* for cleistogamic fertilisation]. If I can trust my memory, the state of the corolla, of the stigma, and the pollen-grains is different from the state of the parts in the bud; that they are in a condition of special modification. But upon my life I am ashamed of myself to differ so much from my betters on this head. The *temporary* theory* which I have formed on this class of dimorphism, just to guide experiment, is that the *perfect* flowers can only be perfectly

* This view is now generally accepted.

fertilised by insects, and are in this case abundantly crossed; but that the flowers are not always, especially in early spring, visited enough by insects, and therefore the little imperfect self-fertilising flowers are developed to ensure a sufficiency of seed for present generations. *Viola canina* is sterile, when not visited by insects, but when so visited forms plenty of seed. I infer from the structure of three or four forms of *Balsamineæ*, that these require insects; at least there is almost as plain adaptation to insects as in Orchids. I have *Oxalis acetosella* ready in pots for experiment next spring; and I fear this will upset my little theory. . . . *Campanula carpathica*, as I found this summer, is absolutely sterile if insects are excluded. *Specularia speculum* is fairly fertile when enclosed; and this seemed to me to be partially effected by the frequent closing of the flower; the inward angular folds of the corolla corresponding with the clefts of the open stigma, and in this action pushing pollen from the outside of the stigma on to its surface. Now can you tell me, does *S. perfoliata* close its flower like *S. speculum*, with angular inward folds? if so, I am smashed without some fearful "wriggling." Are the *imperfect* flowers of your Specularia the early or the later ones? very early or very late? It is rather pretty to see the importance of the closing of flowers of *S. speculum*.

['Forms of Flowers' was published in July 1877; in June he wrote to Professor Carus with regard to the translation:—

"My new book is not a long one, viz. 350 pages, chiefly of the larger type, with fifteen simple woodcuts. All the proofs are corrected except the Index, so that it will soon be published.

". . . . I do not suppose that I shall publish any more books, though perhaps a few more papers. I cannot endure being idle, but heaven knows whether I am capable of any more good work."

The review alluded to in the next letter is at p. 445 of the volume of 'Nature' for 1878:]

C. Darwin to W. Thiselton Dyer.

Down, April 5, 1878.

MY DEAR DYER,—I have just read in 'Nature' the review of 'Forms of Flowers,' and I am sure that it is by you. I wish with all my heart that it deserved one quarter of the praises which you give it. Some of your remarks have interested me greatly. . . . Hearty thanks for your generous and most kind sympathy, which does a man real good, when he is as dog-tired as I am at this minute with working all day, so good-bye.

C. DARWIN.

CHAPTER X.

CLIMBING AND INSECTIVOROUS PLANTS.

[MY father mentions in his 'Autobiography' (vol. i. p. 92) that he was led to take up the subject of climbing plants by reading Dr. Gray's paper, "Note on the Coiling of the Tendrils of Plants." * This essay seems to have been read in 1862, but I am only able to guess at the date of the letter in which he asks for a reference to it, so that the precise date of his beginning this work cannot be determined.

In June 1863 he was certainly at work, and wrote to Sir J. D. Hooker for information as to previous publications on the subject, being then in ignorance of Palm's and H. v. Mohl's works on climbing plants, both of which were published in 1827.]

C. Darwin to J. D. Hooker.

Down [June] 25 [1863].

MY DEAR HOOKER,—I have been observing pretty carefully a little fact which has surprised me; and I want to know from you and Oliver whether it seems new or odd to you, so just tell me whenever you write; it is a very trifling fact, so do not answer on purpose.

I have got a plant of *Echinocystis lobata* to observe the irritability of the tendrils described by Asa Gray, and which of course, is plain enough. Having the plant in my study, I have been surprised to find that the uppermost part of each

* 'Proc. Amer. Acad. of Arts and Sciences,' 1858.

branch (*i.e.* the stem between the two uppermost leaves excluding the growing tip) is *constantly* and slowly twisting round making a circle in from one and a half to two hours; it will sometimes go round two or three times, and then at the same rate untwists and twists in opposite directions. It generally rests half an hour before it retrogrades. The stem does not become permanently twisted. The stem beneath the twisting portion does not move in the least, though not tied. The movement goes on all day and all early night. It has no relation to light, for the plant stands in my window and twists from the light just as quickly as towards it. This may be a common phenomenon for what I know, but it confounded me quite, when I began to observe the irritability of the tendrils. I do not say it is the final cause, but the result is pretty, for the plant every one and a half or two hours sweeps a circle (according to the length of the bending shoot and the length of the tendril) of from one foot to twenty inches in diameter, and immediately that the tendril touches any object its sensitiveness causes it immediately to seize it; a clever gardener, my neighbour, who saw the plant on my table last night, said: "I believe, Sir, the tendrils can see, for wherever I put a plant it finds out any stick near enough." I believe the above is the explanation, viz. that it sweeps slowly round and round. The tendrils have some sense, for they do not grasp each other when young.

Yours affectionately,

C. DARWIN.

C. Darwin to J. D. Hooker.

Down, July 14 [1863].

MY DEAR HOOKER,—I am getting very much amused by my tendrils, it is just the sort of niggling work which suits me, and takes up no time and rather rests me whilst writing. So will you just think whether you know any plant, which you could give or lend me, or I could buy, with tendrils, remarkable in any way for development, for odd or peculiar structure, or even for an odd place in natural arrangement. I have seen or can see Cucurbitaceæ, Passion-flower, Virginian-creeper, *Cissus discolor*, Common-pea and Everlasting-pea. It is really curious the diversification of irritability (I do not mean the spontaneous movement, about which I wrote before and correctly, as further observation shows); for instance, I find a slight pinch between the thumb and finger at the end of the tendril of the Cucurbitaceæ causes prompt movement, but a pinch excites no movement in Cissus. The cause is that one side alone (the concave) is irritable in the former; whereas both sides are irritable in Cissus, so if you excite at the same time both *opposite* sides there is no movement, but by touching with a pencil the two branches of the tendril, in any part whatever, you cause movement towards that point; so that I can mould, by a mere touch, the two branches into any shape I like. . . .

C. Darwin to Asa Gray.

Down, August 4 [1863].

My present hobby-horse I owe to you, viz. the tendrils: their irritability is beautiful, as beautiful in all its modifications as anything in Orchids. About the *spontaneous* movement (independent of touch) of the tendrils and upper internodes, I am rather taken aback by your saying, "is it not well known?" I can find nothing in any book which I have. . . . The spontaneous movement of the tendrils is independent of the movement of the upper internodes, but both work harmoniously together in sweeping a circle for the tendrils to grasp a stick. So with all climbing plants (without tendrils) as yet examined, the upper internodes go on night and day sweeping a circle in one fixed direction. It is surprising to watch the Apocyneæ with shoots 18 inches long (beyond the supporting stick), steadily searching for something to climb up. When the shoot meets a stick, the motion at that point is arrested, but in the upper part is continued ; so that the climbing of all plants yet examined is the simple result of the spontaneous circulatory movement of the upper internodes. Pray tell me whether anything has been published on this subject? I hate publishing what is old ; but I shall hardly regret my work if it is old, as it has much amused me. . . .

C. Darwin to Asa Gray.

May 28, 1864.

. . . . An Irish nobleman on his death-bed declared that he could conscientiously say that he had never throughout life denied himself any pleasure ; and I can conscientiously say that I have never scrupled to trouble you ; so here goes.—Have you travelled South, and can you tell me whether the trees, which *Bignonia capreolata* climbs, are covered with moss or filamentous lichen or Tillandsia? * I ask because its tendrils abhor a simple stick, do not much relish rough bark, but delight in wool or moss. They adhere in a curious manner by making little disks, like the Ampelopsis. . . . By the way, I will enclose some specimens, and if you think it worth while, you can put them under the simple microscope. It is remarkable how specially adapted some tendrils are ; those of *Eccremocarpus scaber* do not like a stick, will have nothing to say to wool ; but give them a bundle of culms of grass, or a bundle of bristles and they seize them well.

C. Darwin to J. D. Hooker.

Down, June 10 [1864].

. . . I have now read two German books, and all I believe that has been written on climbers, and it has stirred me up to

* He subsequently learned from Dr. Gray that *Polypodium incanum* abounds on the trees in the districts where this species of Bignonia grows. See 'Climbing Plants,' p. 103.

find that I have a good deal of new matter. It is strange, but I really think no one has explained simple twining plants. These books have stirred me up, and made me wish for plants specified in them. I shall be very glad of those you mention. I have written to Veitch for young Nepenthes and Vanilla (which I believe will turn out a grand case, though a root creeper), and if I cannot buy young Vanilla I will ask you. I have ordered a leaf-climbing fern, Lygodium. All this work about climbers would hurt my conscience, did I think I could do harder work.*

[He continued his observations on climbing plants during the prolonged illness from which he suffered in the autumn of 1863, and in the following spring. He wrote to Sir J. D. Hooker, apparently in March 1864 :—

"For several days I have been decidedly better, and what I lay much stress on (whatever doctors say), my brain feels far stronger, and I have lost many dreadful sensations. The hot-house is such an amusement to me, and my amusement I owe to you, as my delight is to look at the many odd leaves and plants from Kew. . . . The only approach to work which I can do is to look at tendrils and climbers, this does not distress my weakened brain. Ask Oliver to look over the enclosed queries (and do you look) and amuse a broken-down brother naturalist by answering any which he can. If you ever lounge through your houses, remember me and climbing plants."

On October 29, 1864, he wrote to Dr. Gray :—

"I have not been able to resist doing a little more at your godchild, my climbing paper, or rather in size little book, which by Jove I will have copied out, else I shall never stop. This has been new sort of work for me, and I have been pleased to find what a capital guide for observations a full conviction of the change of species is."

On Jan. 19, 1865, he wrote to Sir J. D. Hooker :—

* He was much out of health at this time.

"It is working hours, but I am trying to take a day's holiday, for I finished and despatched yesterday my climbing paper. For the last ten days I have done nothing but correct refractory sentences, and I loathe the whole subject."

A letter to Dr. Gray, April 9, 1865, has a word or two on the subject.—

"I have begun correcting proofs of my paper on 'Climbing Plants.' I suppose I shall be able to send you a copy in four or five weeks. I think it contains a good deal new and some curious points, but it is so fearfully long, that no one will ever read it. If, however, you do not *skim* through it, you will be an unnatural parent, for it is your child."

Dr. Gray not only read it but approved of it, to my father's great satisfaction, as the following extracts show:—

"I was much pleased to get your letter of July 24th. Now that I can do nothing, I maunder over old subjects, and your approbation of my climbing paper gives me *very* great satisfaction. I made my observations when I could do nothing else and much enjoyed it, but always doubted whether they were worth publishing. I demur to its not being necessary to explain in detail about the spires in *caught* tendrils running in opposite directions; for the fact for a long time confounded me, and I have found it difficult enough to explain the cause to two or three persons." (Aug. 15, 1865.)

"I received yesterday your article * on climbers, and it has pleased me in an extraordinary and even silly manner. You pay me a superb compliment, and as I have just said to my wife, I think my friends must perceive that I like praise, they give me such hearty doses. I always admire your skill in reviews or abstracts, and you have done this article excellently and given the whole essence of my paper. . . . I have had a letter from a good Zoologist in S. Brazil, F. Müller, who has been stirred up to observe climbers and

* In the September number of 'Silliman's Journal,' concluded in the January number, 1866.

gives me some curious cases of *branch*-climbers, in which branches are converted into tendrils, and then continue to grow and throw out leaves and new branches, and then lose their tendril character." (October 1865.)

The paper on Climbing Plants was republished in 1875, as a separate book. The author had been unable to give his customary amount of care to the style of the original essay, owing to the fact that it was written during a period of continued ill-health, and it was now found to require a great deal of alteration. He wrote to Sir J. D. Hooker (March 3, 1875): "It is lucky for authors in general that they do not require such dreadful work in merely licking what they write into shape." And to Mr. Murray in September he wrote: "The corrections are heavy in 'Climbing Plants,' and yet I deliberately went over the MS. and old sheets three times." The book was published in September 1875, an edition of 1500 copies was struck off; the edition sold fairly well, and 500 additional copies were printed in June of the following year.]

INSECTIVOROUS PLANTS.

[In the summer of 1860 he was staying at the house of his sister-in-law, Miss Wedgwood, in Ashdown Forest, whence he wrote (July 29, 1860), to Sir Joseph Hooker:—

"Latterly I have done nothing here; but at first I amused myself with a few observations on the insect-catching power of Drosera; and I must consult you some time whether my 'twaddle' is worth communicating to the Linnean Society."

In August he wrote to the same friend:—

"I will gratefully send my notes on Drosera when copied by my copier: the subject amused me when I had nothing to do."

He has described in the 'Autobiography' (vol. i. p. 95), the general nature of these early experiments. He noticed insects sticking to the leaves, and finding that flies, &c., placed on the adhesive glands were held fast and embraced, he suspected that the leaves were adapted to supply nitrogenous food to the plant. He therefore tried the effect on the leaves of various nitrogenous fluids—with results which, as far as they went, verified his surmise. In September, 1860, he wrote to Dr. Gray:—

"I have been infinitely amused by working at Drosera: the movements are really curious; and the manner in which the leaves detect certain nitrogenous compounds is marvellous. You will laugh; but it is, at present, my full belief (after endless experiments) that they detect (and move in consequence of) the $\frac{1}{2880}$ part of a single grain of nitrate of ammonia; but the muriate and sulphate of ammonia bother their chemical skill, and they cannot make anything of the nitrogen in these salts! I began this work on Drosera in relation to *gradation* as throwing light on Dionæa."

Later in the autumn he was again obliged to leave home for Eastbourne, where he continued his work on Drosera. The work was so new to him that he found himself in difficulties in the preparation of solutions, and became puzzled over fluid and solid ounces, &c. &c. To a friend, the late Mr. E. Cresy, who came to his help in the matter of weights and measures, he wrote giving an account of the experiments. The extract (November 2, 1860) which follows illustrates the almost superstitious precautions he often applied to his researches:—

"Generally I have scrutinised every gland and hair on the leaf before experimenting; but it occurred to me that I might in some way affect the leaf; though this is almost impossible, as I scrutinised with equal care those that I put into distilled water (the same water being used for dissolving the carbonate of ammonia). I then cut off four leaves (not touching them with my fingers), and put them in plain water, and four other leaves into the weak solution, and after leaving them for an hour and a half, I examined every hair on all eight leaves;

no change on the four in water; every gland and hair affected in those in ammonia.

"I had measured the quantity of weak solution, and I counted the glands which had absorbed the ammonia, and were plainly affected; the result convinced me that each gland could not have absorbed more than $\frac{1}{64000}$ or $\frac{1}{65000}$ of a grain. I have tried numbers of other experiments all pointing to the same result. Some experiments lead me to believe that very sensitive leaves are acted on by much smaller doses. Reflect how little ammonia a plant can get growing on poor soil—yet it is nourished. The really surprising part seems to me that the effect should be visible, and not under very high power; for after trying a high power, I thought it would be safer not to consider any effect which was not plainly visible under a two-thirds object glass and middle eye-piece. The effect which the carbonate of ammonia produces is the segregation of the homogeneous fluid in the cells into a cloud of granules and colourless fluid; and subsequently the granules coalesce into larger masses, and for hours have the oddest movements—coalescing, dividing, coalescing *ad infinitum.* I do not know whether you will care for these ill-written details; but, as you asked, I am sure I am bound to comply, after all the very kind and great trouble which you have taken."

On his return home he wrote to Sir J. D. Hooker (November 21, 1860):—

"I have been working like a madman at Drosera. Here is a fact for you which is certain as you stand where you are, though you won't believe it, that a bit of hair $\frac{1}{78000}$ of one grain in weight placed on gland, will cause *one* of the gland-bearing hairs of Drosera to curve inwards, and will alter the condition of the contents of every cell in the foot-stalk of the gland."

And a few days later to Lyell:—

"I will and must finish my Drosera MS., which will take

me a week, for, at the present moment, I care more about Drosera than the origin of all the species in the world. But I will not publish on Drosera till next year, for I am frightened and astounded at my results. I declare it is a certain fact, that one organ is so sensitive to touch, that a weight seventy-eight-times less than that, viz., $\frac{1}{1000}$ of a grain, which will move the best chemical balance, suffices to cause a conspicuous movement. Is it not curious that a plant should be far more sensitive to the touch than any nerve in the human body? Yet I am perfectly sure that this is true. When I am on my hobby-horse, I never can resist telling my friends how well my hobby goes, so you must forgive the rider."

The work was continued, as a holiday task, at Bournemouth, where he stayed during the autumn of 1862. The discussion in the following letter on "nervous matter" in Drosera is of interest in relation to recent researches on the continuity of protoplasm from cell to cell :]

C. Darwin to J. D. Hooker.

Cliff Cottage, Bournemouth.
September 26 [1862].

MY DEAR HOOKER,—Do not read this till you have leisure. If that blessed moment ever comes, I should be very glad to have your opinion on the subject of this letter. I am led to the opinion that Drosera must have diffused matter in organic connection, closely analogous to the nervous matter of animals. When the glands of one of the papillæ or tentacles, in its natural position is supplied with nitrogenised fluid and certain other stimulants, or when loaded with an extremely slight weight, or when struck several times with a needle, the pedicel bends near its base in under one minute. These varied stimulants are conveyed down the pedicel by some means; it cannot be vibration, for drops of fluid put on quite quietly cause the movement; it cannot be absorption of the fluid from cell to cell, for I can see the rate of absorption, which though quick, is far slower, and in Dionæa the transmission is instantaneous; analogy from animals would point to transmission through nervous matter. Reflecting on the rapid power of absorption in the glands, the extreme sensibility of the whole organ, and the conspicuous movement caused by varied stimulants, I have tried a number of substances which are not caustic or corrosive, ... but most of which are known to have a remarkable action on the nervous matter of animals. You will see the results in the enclosed paper. As the nervous matter of different animals are differently acted on by the same poisons, one would not expect the same action on plants and animals; only, if plants have diffused nervous matter, some degree of analogous action. And this is partially the case. Considering these experiments, together with the previously made remarks on the functions of the parts, I cannot avoid the conclusion, that Drosera possesses matter at least in some degree analogous in constitution and function to nervous matter. Now do tell me what you think, as far as you can judge from my abstract; of course many more experiments would have to be tried; but in former years I tried on the whole leaf, instead of on separate glands, a number of innocuous * substances, such as sugar, gum, starch, &c., and they produced no effect. Your opinion will aid me in deciding some future year in going on with this subject. I should not have thought it worth attempting, but I had nothing on earth to do.

My dear Hooker, yours very sincerely,

CH. DARWIN.

P.S.—We return home on Monday 28th. Thank Heaven!

* This line of investigation made him wish for information on the action of poisons on plants; as in many other cases he applied to Professor Oliver, and in reference to the result wrote to Hooker: "Pray thank Oliver heartily for his heap of references on poisons."

[A long break now ensued in his work on insectivorous plants, and it was not till 1872 that the subject seriously occupied him again. A passage in a letter to Dr. Asa Gray, written in 1863 or 1864, shows, however, that the question was not altogether absent from his mind in the interim:—

"Depend on it you are unjust on the merits of my beloved Drosera; it is a wonderful plant, or rather a most sagacious animal. I will stick up for Drosera to the day of my death. Heaven knows whether I shall ever publish my pile of experiments on it."

He notes in his diary that the last proof of the 'Expression of the Emotions' was finished on August 22, 1872, and that he began to work on Drosera on the following day.]

C. Darwin to Asa Gray.

[Sevenoaks], October 22 [1872].

. . . I have worked pretty hard for four or five weeks on Drosera, and then broke down; so that we took a house near Sevenoaks for three weeks (where I now am) to get complete rest. I have very little power of working now, and must put off the rest of the work on Drosera till next spring, as my plants are dying. It is an endless subject, and I must cut it short, and for this reason shall not do much on Dionæa. The point which has interested me most is tracing the *nerves!* which follow the vascular bundles. By a prick with a sharp lancet at a certain point, I can paralyse one-half the leaf, so that a stimulus to the other half causes no movement. It is just like dividing the spinal marrow of a frog:—no stimulus can be sent from the brain or anterior part of the spine to the hind legs; but if these latter are stimulated, they move by reflex action. I find my old results about the astonishing sensitiveness of the nervous system (!?) of Drosera to various stimulants fully confirmed and extended. . . .

[His work on digestion in Drosera and on other points in

the physiology of the plant soon led him into regions where his knowledge was defective, and here the advice and assistance which he received from Dr. Burdon Sanderson was of much value :]

C. Darwin to J. Burdon Sanderson.

Down, July 25, 1873.

My dear Dr. Sanderson,—I should like to tell you a little about my recent work with Drosera, to show that I have profited by your suggestions, and to ask a question or two.

1. It is really beautiful how quickly and well Drosera and Dionæa dissolve little cubes of albumen and gelatine. I kept the same sized cubes on wet moss for comparison. When you were here I forgot that I had tried gelatine, but albumen is far better for watching its dissolution and absorption. Frankland has told me how to test in a rough way for pepsine; and in the autumn he will discover what acid the digestive juice contains.

2. A decoction of cabbage-leaves and green peas causes as much inflection as an infusion of raw meat; a decoction of grass is less powerful. Though I hear that the chemists try to precipitate all albumen from the extract of belladonna, I think they must fail, as the extract causes inflection, whereas a new lot of atropine, as well as the valerianate [of atropine], produce no effect.

3. I have been trying a good many experiments with heated water. . . . Should you not call the following case one of heat rigor? Two leaves were heated to 130°, and had every tentacle closely inflected; one was taken out and placed in cold water, and it re-expanded; the other was heated to 145°, and had not the least power of re-expansion. Is not this latter case heat rigor? If you can inform me, I should very much like to hear at what temperature cold-blooded and invertebrate animals are killed.

4. I must tell you my final result, of which I am sure, [as to] the sensitiveness of Drosera. I made a solution of one part of phosphate of ammonia by weight to 218,750 of water; of this solution I gave so much that a leaf got $\frac{1}{8000}$ of a grain of the phosphate. I then counted the glands, and each could have got only $\frac{1}{1552000}$ of a grain; this being absorbed by the glands, sufficed to cause the tentacles bearing these glands to bend through an angle of 180°. Such sensitiveness requires hot weather, and carefully selected young yet mature leaves. It strikes me as a wonderful fact. I must add that I took every precaution, by trying numerous leaves at the same time in the solution and in the same water which was used for making the solution.

5. If you can persuade your friend to try the effects of carbonate of ammonia on the aggregation of the white blood corpuscles, I should very much like to hear the result.

I hope this letter will not have wearied you.

Believe me, yours very sincerely,

CHARLES DARWIN.

C. Darwin to W. Thiselton Dyer.

Down, 24 [December 1873 ?].

MY DEAR MR. DYER,—I fear that you will think me a great bore, but I cannot resist telling you that I have just found out that the leaves of Pinguicula possess a beautifully adapted power of movement. Last night I put on a row of little flies near one edge of two *youngish* leaves; and after 14 hours these edges are beautifully folded over so as to clasp the flies, thus bringing the glands into contact with the upper surfaces of the flies, and they are now secreting copiously above and below the flies and no doubt absorbing. The acid secretion has run down the channelled edge and has collected in the spoon-shaped extremity, where no doubt the glands are absorbing the delicious soup. The leaf on one side looks just like the helix of a human ear, if you were to stuff flies within the fold.

Yours most sincerely,
CH. DARWIN.

C. Darwin to Asa Gray.

Down, June 3 [1874].

. . . . I am now hard at work getting my book on Drosera & Co. ready for the printers, but it will take some time, for I am always finding out new points to observe. I think you will be interested by my observations on the digestive process in Drosera; the secretion contains an acid of the acetic series, and some ferment closely analogous to, but not identical with, pepsine; for I have been making a long series of comparative trials. No human being will believe what I shall publish about the smallness of the doses of phosphate of ammonia which act.

. . . . I began reading the Madagascar squib * quite gravely, and when I found it stated that Felis and Bos inhabited Madagascar, I thought it was a false story, and did not perceive it was a hoax till I came to the woman. . . .

C. Darwin to F. C. Donders.†

Down, July 7, 1874.

MY DEAR PROFESSOR DONDERS,—My son George writes to me that he has seen you, and that you have been very kind to him, for which I return to you my cordial thanks. He tells me on your authority, of a fact which interests me in the highest degree, and which I much wish to be allowed to quote. It relates to the action of one millionth of a grain of atropine on the eye. Now will you be so kind, whenever you can find a little leisure, to tell me whether you yourself have

* A description of a carnivorous plant supposed to subsist on human beings.

† Professor Donders, the well-known physiologist of Utrecht.

observed this fact, or believe it on good authority. I also wish to know what proportion by weight the atropine bore to the water of solution, and how much of the solution was applied to the eye. The reason why I am so anxious on this head is that it gives some support to certain facts repeatedly observed by me with respect to the action of phosphate of ammonia on Drosera. The $\frac{1}{4000000}$ of a grain absorbed by a gland clearly makes the tentacle which bears this gland become inflected; and I am fully convinced that $\frac{1}{20000000}$ of a grain of the crystallised salt (*i.e.* containing about one-third of its weight of water of crystallisation) does the same. Now I am quite unhappy at the thought of having to publish such a statement. It will be of great value to me to be able to give any analogous facts in support. The case of Drosera is all the more interesting as the absorption of the salt or any other stimulant applied to the gland causes it to transmit a motor influence to the base of the tentacle which bears the gland.

Pray forgive me for troubling you, and do not trouble yourself to answer this until your health is fully re-established.

Pray believe me,

Yours very sincerely,

CHARLES DARWIN.

[During the summer of 1874 he was at work on the genus Utricularia, and he wrote (July 16th) to Sir J. D. Hooker giving some account of the progress of his work:—

"I am rather glad you have not been able to send Utricularia, for the common species has driven F. and me almost mad. The structure is *most* complex. The bladders catch a multitude of Entomostraca, and larvæ of insects. The mechanism for capture is excellent. But there is much that we cannot understand. From what I have seen to-day, strongly suspect that it is necrophagous, *i.e.* that it cannot digest, but absorbs decaying matter."

He was indebted to Lady Dorothy Nevill for specimens of the curious *Utricularia montana*, which is not aquatic like the European species, but grows among the moss and *débris* on the branches of trees. To this species the following letter refers :]

C. Darwin to Lady Dorothy Nevill.

Down, September 18 [1874].

DEAR LADY DOROTHY NEVILL,—I am so much obliged to you. I was so convinced that the bladders were with the leaves that I never thought of removing the moss, and this was very stupid of me. The great solid bladder-like swellings almost on the surface are wonderful objects, but are not the true bladders. These I found on the roots near the surface, and down to a depth of two inches in the sand. They are as transparent as glass, from $\frac{1}{20}$ to $\frac{1}{100}$ of an inch in size, and hollow. They have all the important points of structure of the bladders of the floating English species, and I felt confident I should find captured prey. And so I have to my delight in two bladders, with clear proof that they had absorbed food from the decaying mass. For Utricularia is a carrion-feeder, and not strictly carnivorous like Drosera.

The great solid bladder-like bodies, I believe, are reservoirs of water like a camel's stomach. As soon as I have made a few more observations, I mean to be so cruel as to give your plant no water, and observe whether the great bladders shrink and contain air instead of water; I shall then also wash all earth from all roots, and see whether there are true bladders for capturing subterranean insects down to the very bottom of the pot. Now shall you think me very greedy, if I say that supposing the species is not very precious, and you have several, will you give me one more plant, and if so, please to send it to "Orpington Station, S. E. R., to be forwarded by foot messenger."

I have hardly ever enjoyed a day more in my life than I

have this day's work; and this I owe to your Ladyship's great kindness.

The seeds are very curious monsters; I fancy of some plant allied to Medicago, but I will show them to Dr. Hooker.

Your Ladyship's very gratefully,

CH. DARWIN.

C. Darwin to J. D. Hooker.

Down, September 30, 1874.

MY DEAR H.,—Your magnificent present of Aldrovanda has arrived quite safe. I have enjoyed greatly a good look at the shut leaves, one of which I cut open. It is an aquatic Dionæa, which has acquired some structures identical with those of Utricularia!

If the leaves open, and I can transfer them open under the microscope, I will try some experiments, for mortal man cannot resist the temptation. If I cannot transfer, I will do nothing, for otherwise it would require hundreds of leaves.

You are a good man to give me such pleasure.

Yours affectionately,

C. DARWIN.

[The manuscript of 'Insectivorous Plants' was finished in March 1875. He seems to have been more than usually oppressed by the writing of this book, thus he wrote to Sir J. D. Hooker in February:—

"You ask about my book, and all that I can say is that I am ready to commit suicide; I thought it was decently written, but find so much wants rewriting, that it will not be ready to go to printers for two months, and will then make a confoundedly big book. Murray will say that it is no use publishing in the middle of summer, so I do not know what will be the upshot; but I begin to think that every one who publishes a book is a fool."

The book was published on July 2nd, 1875, and 2700 copies were sold out of the edition of 3000.]

CHAPTER XI.

THE 'POWER OF MOVEMENT IN PLANTS.' 1880.

[THE few sentences in the autobiographical chapter give with sufficient clearness the connection between the 'Power of Movement,' and one of the author's earlier books, that on 'Climbing Plants.' The central idea of the book is that the movements of plants in relation to light, gravitation, &c., are modifications of a spontaneous tendency to revolve or circumnutate, which is widely inherent in the growing parts of plants. This conception has not been generally adopted, and has not taken a place among the canons of orthodox physiology. The book has been treated by Professor Sachs with a few words of professorial contempt; and by Professor Wiesner it has been honoured by careful and generously expressed criticism.

Mr. Thiselton Dyer * has well said: "Whether this masterly conception of the unity of what has hitherto seemed a chaos of unrelated phenomena will be sustained, time alone will show. But no one can doubt the importance of what Mr. Darwin has done, in showing that for the future the phenomena of plant movement can and indeed must be studied from a single point of view."

The work was begun in the summer of 1877, after the publication of 'Different Forms of Flowers,' and by the autumn his enthusiasm for the subject was thoroughly established, and he wrote to Mr. Dyer: "I am all on fire at the

* 'Charles Darwin' ('Nature' Series), p. 41.

work." At this time he was studying the movements of cotyledons, in which the sleep of plants is to be observed in its simplest form; in the following spring he was trying to discover what useful purpose these sleep-movements could serve, and wrote to Sir Joseph Hooker (March 25th, 1878):—

"I think we have *proved* that the sleep of plants is to lessen the injury to the leaves from radiation. This has interested me much, and has cost us great labour, as it has been a problem since the time of Linnæus. But we have killed or badly injured a multitude of plants: N.B.—*Oxalis carnosa* was most valuable, but last night was killed."

His letters of this period do not give any connected account of the progress of the work. The two following seem worth giving as being characteristic of the author:]

C. Darwin to W. Thiselton Dyer.

Down, June 2, 1878.

MY DEAR DYER,—I remember saying that I should die a disgraced man if I did not observe a seedling Cactus and Cycas, and you have saved me from this horrible fate, as they move splendidly and normally. But I have two questions to ask: the Cycas observed was a huge seed in a broad and very shallow pot with cocoa-nut fibre as I suppose. It was named only Cycas. Was it *Cycas pectinata?* I suppose that I cannot be wrong in believing that what first appears above ground is a true leaf, for I can see no stem or axis. Lastly, you may remember that I said that we could not raise *Opuntia nigricans;* now I must confess to a piece of stupidity; one did come up, but my gardener and self stared at it, and concluded that it could not be a seedling Opuntia, but now that I have seen one of *O. basilaris*, I am sure it was; I observed it only casually, and saw movements, which makes me wish to observe carefully another. If you have any fruit, will Mr. Lynch * be so kind as to send one more?

I am working away like a slave at radicles [roots] and at movements of true leaves, for I have pretty well done with cotyledons. . . .

That was an *excellent* letter about the Gardens: † I had hoped that the agitation was over. Politicians are a poor truckling lot, for [they] must see the wretched effects of keeping the gardens open all day long.

Your ever troublesome friend,

CH. DARWIN.

C. Darwin to W. Thiselton Dyer.

4 Bryanston St., Portman Square,
November 21 [1878].

MY DEAR DYER,—I must thank you for all the wonderful trouble which you have taken about the seeds of *Impatiens* and on scores of other occasions. It in truth makes me feel ashamed of myself, and I cannot help thinking: "Oh Lord, when he sees our book he will cry out, is this all for which I have helped so much!" In seriousness, I hope that we have made out some points, but I fear that we have done very little for the labour which we have expended on our work. We are here for a week for a little rest, which I needed.

If I remember right, November 30th, is the anniversary at the Royal, and I fear Sir Joseph must be almost at the last gasp. I shall be glad when he is no longer President.

Yours very sincerely,

CH. DARWIN.

[In the spring of the following year, 1879, when he was engaged in putting his results together, he wrote somewhat

* Mr. R. I. Lynch, now Curator of the Botanic Garden at Cambridge, was at this time in the Royal Gardens, Kew.

† This refers to an attempt to induce the Government to open the Royal Gardens at Kew in the morning.

despondingly to Mr. Dyer: "I am overwhelmed with my notes, and almost too old to undertake the job which I have in hand—*i.e.*, movements of all kinds. Yet it is worse to be idle."

Later on in the year, when the work was approaching completion, he wrote to Prof. Carus (July 17, 1879), with respect to a translation:—

"Together with my son Francis, I am preparing a rather large volume on the general movements of Plants, and I think that we have made out a good many new points and views.

"I fear that our views will meet a good deal of opposition in Germany; but we have been working very hard for some years at the subject.

"I shall be *much* pleased if you think the book worth translating, and proof-sheets shall be sent you, whenever they are ready."

In the autumn he was hard at work on the manuscript, and wrote to Dr. Gray (October 24, 1879):—

"I have written a rather big book—more is the pity—on the movements of plants, and I am now just beginning to go over the MS. for the second time, which is a horrid bore."

Only the concluding part of the next letter refers to the 'Power of Movement':]

C. Darwin to A. De Candolle.

May 28, 1880.

MY DEAR SIR,—I am particularly obliged to you for having so kindly sent me your 'Phytographie;' * for if I had merely seen it advertised, I should not have supposed that it could have concerned me. As it is, I have read with very great interest about a quarter, but will not delay longer thanking you. All that you say seems to me very clear and convincing, and as in all your writings I find a large number of

* A book on the methods of botanical research, more especially of systematic work.

philosophical remarks new to me, and no doubt shall find many more. They have recalled many a puzzle through which I passed when monographing the Cirripedia; and your book in those days would have been quite invaluable to me. It has pleased me to find that I have always followed your plan of making notes on separate pieces of paper; I keep several scores of large portfolios, arranged on very thin shelves about two inches apart, fastened to the walls of my study, and each shelf has its proper name or title; and I can thus put at once every memorandum into its proper place. Your book will, I am sure, be very useful to many young students, and I shall beg my son Francis (who intends to devote himself to the physiology of plants) to read it carefully.

As for myself I am taking a fortnight's rest, after sending a pile of MS. to the printers, and it was a piece of good fortune that your book arrived as I was getting into my carriage, for I wanted something to read whilst away from home. My MS. relates to the movements of plants, and I think that I have succeeded in showing that all the more important great classes of movements are due to the modification of a kind of movement common to all parts of all plants from their earliest youth.

Pray give my kind remembrances to your son, and with my highest respect and best thanks,

Believe me, my dear Sir, yours very sincerely,

CHARLES DARWIN.

P.S.—It always pleases me to exalt plants in the organic scale, and if you will take the trouble to read my last chapter when my book (which will be sadly too big) is published and sent to you, I hope and think that you also will admire some of the beautiful adaptations by which seedling plants are enabled to perform their proper functions.

[The book was published on November 6, 1880, and 1500

copies were disposed of at Mr. Murray's sale. With regard to it he wrote to Sir J. D. Hooker (November 23):—

"Your note has pleased me much—for I did not expect that you would have had time to read *any* of it. Read the last chapter, and you will know the whole result, but without the evidence. The case, however, of radicles bending after exposure for an hour to geotropism, with their tips (or brains) cut off is, I think, worth your reading (bottom of p. 525); it astounded me. The next most remarkable fact, as it appeared to me (p. 148), is the discrimination of the tip of the radicle between a slightly harder and softer object affixed on opposite sides of tip. But I will bother you no more about my book. The sensitiveness of seedlings to light is marvellous."

To another friend, Mr. Thiselton Dyer, he wrote (November 28, 1880):—

"Very many thanks for your most kind note, but you think too highly of our work, not but what this is very pleasant. . . . Many of the Germans are very contemptuous about making out the use of organs; but they may sneer the souls out of their bodies, and I for one shall think it the most interesting part of Natural History. Indeed you are greatly mistaken if you doubt for one moment on the very great value of your constant and most kind assistance to us."

The book was widely reviewed, and excited much interest among the general public. The following letter refers to a leading article in the *Times*, November 20, 1880:]

*C. Darwin to Mrs. Haliburton.**

Down, November 22, 1880.

MY DEAR SARAH,—You see how audaciously I begin; but I have always loved and shall ever love this name. Your

* Mrs. Haliburton is a daughter of my father's early friend, the late Mr. Owen, of Woodhouse.

letter has done more than please me, for its kindness has touched my heart. I often think of old days and of the delight of my visits to Woodhouse, and of the deep debt of gratitude which I owe to your father. It was very good of you to write. I had quite forgotten my old ambition about the Shrewsbury newspaper;* but I remember the pride which I felt when I saw in a book about beetles the impressive words "captured by C. Darwin." Captured sounded so grand compared with caught. This seemed to me glory enough for any man! I do not know in the least what made the *Times* glorify me,† for it has sometimes pitched into me ferociously.

I should very much like to see you again, but you would find a visit here very dull, for we feel very old and have no amusement, and lead a solitary life. But we intend in a few weeks to spend a few days in London, and then if you have anything else to do in London, you would perhaps come and lunch with us.‡

Believe me, my dear Sarah,

Yours gratefully and affectionately,

CHARLES DARWIN.

[The following letter was called forth by the publication of a volume devoted to the criticism of the 'Power of Movement in Plants' by an accomplished botanist, Dr. Julius Wiesner, Professor of Botany in the University of Vienna:]

* Mrs. Haliburton had reminded him of his saying as a boy that if Eddowes' newspaper ever alluded to him as "our deserving fellow-townsman," his ambition would be amply gratified.

† The following is the opening sentence of the leading article:—"Of all our living men of science none have laboured longer and to more splendid purpose than Mr. Darwin."

‡ My father had the pleasure of seeing Mrs. Haliburton at his brother's house in Queen Anne Street.

C. Darwin to Julius Wiesner.

Down, October 25th, 1881.

MY DEAR SIR,—I have now finished your book,* and have understood the whole except a very few passages. In the first place, let me thank you cordially for the manner in which you have everywhere treated me. You have shown how a man may differ from another in the most decided manner, and yet express his difference with the most perfect courtesy. Not a few English and German naturalists might learn a useful lesson from your example; for the coarse language often used by scientific men towards each other does no good, and only degrades science.

I have been profoundly interested by your book, and some of your experiments are so beautiful, that I actually felt pleasure while being vivisected. It would take up too much space to discuss all the important topics in your book. I fear that you have quite upset the interpretation which I have given of the effects of cutting off the tips of horizontally extended roots, and of those laterally exposed to moisture; but I cannot persuade myself that the horizontal position of lateral branches and roots is due simply to their lessened power of growth. Nor when I think of my experiments with the cotyledons of *Phalaris*, can I give up the belief of the transmission of some stimulus due to light from the upper to the lower part. At p. 60 you have misunderstood my meaning, when you say that I believe that the effects from light are transmitted to a part which is not itself heliotropic. I never considered whether or not the short part beneath the ground was heliotropic; but I believe that with young seedlings the part which bends *near*, but *above* the ground is heliotropic, and I believe so from this part bending only moderately when the light is oblique, and bending rectangularly when the light is horizontal. Nevertheless the bending

* 'Das Bewegungsvermögen der Pflanzen.' Vienna, 1881.

of this lower part, as I conclude from my experiments with opaque caps, is influenced by the action of light on the upper part. My opinion, however, on the above and many other points, signifies very little, for I have no doubt that your book will convince most botanists that I am wrong in all the points on which we differ.

Independently of the question of transmission, my mind is so full of facts leading me to believe that light, gravity, &c., act not in a direct manner on growth, but as stimuli, that I am quite unable to modify my judgment on this head. I could not understand the passage at p. 78, until I consulted my son George, who is a mathematician. He supposes that your objection is founded on the diffused light from the lamp illuminating both sides of the object, and not being reduced, with increasing distance in the same ratio as the direct light; but he doubts whether this *necessary* correction will account for the very little difference in the heliotropic curvature of the plants in the successive pots.

With respect to the sensitiveness of the tips of roots to contact, I cannot admit your view until it is proved that I am in error about bits of card attached by liquid gum causing movement; whereas no movement was caused if the card remained separated from the tip by a layer of the liquid gum. The fact also of thicker and thinner bits of card attached on opposite sides of the same root by shellac, causing movement in one direction, has to be explained. You often speak of the tip having been injured; but externally there was no sign of injury: and when the tip was plainly injured, the extreme part became curved *towards* the injured side. I can no more believe that the tip was injured by the bits of card, at least when attached by gum-water, than that the glands of Drosera are injured by a particle of thread or hair placed on it, or that the human tongue is so when it feels any such object.

About the most important subject in my book, namely circumnutation, I can only say that I feel utterly bewildered at the difference in our conclusions; but I could not fully understand some parts which my son Francis will be able to translate to me when he returns home. The greater part of your book is beautifully clear.

Finally, I wish that I had enough strength and spirit to commence a fresh set of experiments, and publish the results, with a full recantation of my errors when convinced of them; but I am too old for such an undertaking, nor do I suppose that I shall be able to do much, or any more, original work. I imagine that I see one possible source of error in your beautiful experiment of a plant rotating and exposed to a lateral light.

With high respect and with sincere thanks for the kind manner in which you have treated me and my mistakes, I remain,

My dear Sir, yours sincerely,

CHARLES DARWIN.

CHAPTER XII.

MISCELLANEOUS BOTANICAL LETTERS.

1873–1882.

[THE present chapter contains a series of miscellaneous letters on botanical subjects. Some of them show my father's varied interests in botanical science, and others give account of researches which never reached completion.]

BLOOM ON LEAVES AND FRUIT.

[His researches into the meaning of the "bloom," or waxy coating found on many leaves, was one of those inquiries which remained unfinished at the time of his death. He amassed a quantity of notes on the subject, part of which I hope to publish at no distant date.*

One of his earliest letters on this subject was addressed in August, 1873, to Sir Joseph Hooker:—

"I want a little information from you, and if you do not yourself know, please to enquire of some of the wise men of Kew.

"Why are the leaves and fruit of so many plants protected by a thin layer of waxy matter (like the common cabbage),

* A small instalment, on the relation between bloom and the distribution of the stomata on leaves, has appeared in the 'Journal of the Linnean Society,' 1886. Tschirsch (*Linnæa*, 1881) has published results identical with some which my father and myself obtained, viz. that bloom diminishes transpiration. The same fact was previously published by Garreau, in 1850.

or with fine hair, so that when such leaves or fruit are immersed in water they appear as if encased in thin glass? It is really a pretty sight to put a pod of the common pea, or a raspberry into water. I find several leaves are thus protected on the under surface and not on the upper.

"How can water injure the leaves if indeed this is at all the case?"

On this latter point he wrote to Sir Thomas Farrer:—

"I am now become mad about drops of water injuring leaves. Please ask Mr. Paine * whether he believes, *from his own experience*, that drops of water injure leaves or fruit in his conservatories. It is said that the drops act as burning-glasses; if this is true, they would not be at all injurious on cloudy days. As he is so acute a man, I should very much like to hear his opinion. I remember when I grew hot-house orchids I was cautioned not to wet their leaves; but I never then thought on the subject.

"I enjoyed my visit greatly with you, and I am very sure that all England could not afford a kinder and pleasanter host."

Some years later he took up the subject again, and wrote to Sir Joseph Hooker (May 25, 1877):—

"I have been looking over my old notes about the "bloom" on plants, and I think that the subject is well worth pursuing, though I am very doubtful of any success. Are you inclined to aid me on the mere chance of success, for without your aid I could do hardly anything?"]

C. Darwin to Asa Gray.

Down, June 4 [1877].

. . . . I am now trying to make out the use or function of "bloom," or the waxy secretion on the leaves and fruit of plants, but am *very* doubtful whether I shall succeed. Can

* Sir Thomas Farrer's gardener.

you give me any light? Are such plants commoner in warm than in colder climates? I ask because I often walk out in heavy rain, and the leaves of very few wild dicotyledons can be here seen with drops of water rolling off them like quicksilver. Whereas in my flower garden, greenhouse, and hot-houses there are several. Again, are bloom-protected plants common on your *dry* western plains? Hooker *thinks* that they are common at the Cape of Good Hope. It is a puzzle to me if they are common under very dry climates, and I find bloom very common on the Acacias and Eucalypti of Australia. Some of the Eucalypti which do not appear to be covered with bloom have the epidermis protected by a layer of some substance which is dissolved in boiling alcohol. Are there any bloom-protected leaves or fruit in the Arctic regions? If you can illuminate me, as you so often have done, pray do so; but otherwise do not bother yourself by answering.

Yours affectionately,

C. DARWIN.

C. Darwin to W. Thiselton Dyer.

Down, September 5 [1877].

MY DEAR DYER,—One word to thank you. I declare had it not been for your kindness, we should have broken down. As it is we have made out clearly that with some plants (chiefly succulent) the bloom checks evaporation—with some certainly prevents attacks of insects; with *some* sea-shore plants prevents injury from salt-water, and, I believe, with a few prevents injury from pure water resting on the leaves. This latter is as yet the most doubtful and the most interesting point in relation to the movements of plants.

C. Darwin to F. Müller.

Down, July 4 [1881].

My dear Sir,—Your kindness is unbounded, and I cannot tell you how much your last letter (May 31) has interested me. I have piles of notes about the effect of water resting on leaves, and their movements (as I supposed) to shake off the drops. But I have not looked over these notes for a long time, and had come to think that perhaps my notion was mere fancy, but I had intended to begin experimenting as soon as I returned home; and now with your *invaluable* letter about the position of the leaves of various plants during rain (I have one analogous case with Acacia from South Africa), I shall be stimulated to work in earnest.

Variability.

[The following letter refers to a subject on which my father felt the strongest interest:—the experimental investigation of the causes of variability. The experiments alluded to were to some extent planned out, and some preliminary work was begun in the direction indicated below, but the research was ultimately abandoned.]

*C. Darwin to J. H. Gilbert.**

Down, February 16, 1876.

My dear Sir,—When I met you at the Linnean Society, you were so kind as to say that you would aid me with advice, and this will be of the utmost value to me and my son. I will first state my object, and hope that you will excuse a long letter. It is admitted by all naturalists that no problem is so perplexing as what causes almost every cultivated plant to

* Dr. Gilbert, F.R.S., joint author with Sir John Bennett Lawes of a long series of valuable researches in Scientific Agriculture.

vary, and no experiments as yet tried have thrown any light on the subject. Now for the last ten years I have been experimenting in crossing and self-fertilising plants; and one indirect result has surprised me much; namely, that by taking pains to cultivate plants in pots under glass during several successive generations, under nearly similar conditions, and by self-fertilising them in each generation, the colour of the flowers often changes, and, what is very remarkable, they became in some of the most variable species, such as Mimulus, Carnation, &c., quite constant, like those of a wild species.

This fact and several others have led me to the suspicion that the cause of variation must be in different substances absorbed from the soil by these plants when their powers of absorption are not interfered with by other plants with which they grow mingled in a state of nature. Therefore my son and I wish to grow plants in pots in soil entirely, or as nearly entirely as is possible, destitute of all matter which plants absorb, and then to give during several successive generations to several plants of the same species as different solutions as may be compatible with their life and health. And now, can you advise me how to make soil approximately free of all the substances which plants naturally absorb? I suppose white silver sand, sold for cleaning harness, &c., is nearly pure silica, but what am I to do for alumina? Without some alumina I imagine that it would be impossible to keep the soil damp and fit for the growth of plants. I presume that clay washed over and over again in water would still yield mineral matter to the carbonic acid secreted by the roots. I should want a good deal of soil, for it would be useless to experimentise unless we could fill from twenty to thirty moderately sized flower-pots every year. Can you suggest any plan? for unless you can it would, I fear, be useless for us to commence an attempt to discover whether variability depends at all on matter absorbed from the soil. After obtaining the requisite kind of soil, my notion is to water one set of plants with nitrate of potassium, another set with nitrate of sodium, and another with nitrate of lime, giving all as much phosphate of ammonia as they seemed to support, for I wish the plants to grow as luxuriantly as possible. The plants watered with nitrate of Na and of Ca would require, I suppose, some K; but perhaps they would get what is absolutely necessary from such soil as I should be forced to employ, and from the rain-water collected in tanks. I could use hard water from a deep well in the chalk, but then all the plants would get lime. If the plants to which I give Nitrate of Na and of Ca would not grow I might give them a little alum.

I am well aware how very ignorant I am, and how crude my notions are; and if you could suggest any other solutions by which plants would be likely to be affected it would be a very great kindness. I suppose that there are no organic fluids which plants would absorb, and which I could procure?

I must trust to your kindness to excuse me for troubling you at such length, and,

I remain, dear Sir, yours sincerely,

CHARLES DARWIN.

[The next letter to Professor Semper bears on the same subject:]

*From C. Darwin to K. Semper.**

Down, July 19, 1881.

MY DEAR PROFESSOR SEMPER,—I have been much pleased to receive your letter, but I did not expect you to answer my former one. . . . . I cannot remember what I wrote to you, but I am sure that it must have expressed the interest which I felt in reading your book.† I thought that you attributed too much weight to the *direct* action of the

* Professor of Zoology at Würzburg.

† Published in the 'International Scientific Series,' in 1881, under the title, 'The Natural Conditions of Existence as they affect Animal Life.'

environment; but whether I said so I know not, for without being asked I should have thought it presumptuous to have criticised your book, nor should I now say so had I not during the last few days been struck with Professor Hoffmann's review of his own work in the 'Botanische Zeitung,' on the variability of plants; and it is really surprising how little effect he produced by cultivating certain plants under unnatural conditions, as the presence of salt, lime, zinc, &c., &c., during *several* generations. Plants, moreover, were selected which were the most likely to vary under such conditions, judging from the existence of closely-allied forms adapted for these conditions. No doubt I originally attributed too little weight to the direct action of conditions, but Hoffmann's paper has staggered me. Perhaps hundreds of generations of exposure are necessary. It is a most perplexing subject. I wish I was not so old, and had more strength, for I see lines of research to follow. Hoffmann even doubts whether plants vary more under cultivation than in their native home and under their natural conditions. If so, the astonishing variations of almost all cultivated plants must be due to selection and breeding from the varying individuals. This idea crossed my mind many years ago, but I was afraid to publish it, as I thought that people would say, "how he does exaggerate the importance of selection."

I still *must* believe that changed conditions give the impulse to variability, but that they act *in most cases* in a very indirect manner. But, as I said, it is a most perplexing problem. Pray forgive me for writing at such length; I had no intention of doing so when I sat down to write.

I am extremely sorry to hear, for your own sake and for that of Science, that you are so hard worked, and that so much of your time is consumed in official labour.

Pray believe me, dear Professor Semper,

Yours sincerely,

CHARLES DARWIN.

GALLS.

[Shortly before his death, my father began to experimentise on the possibility of producing galls artificially. A letter to Sir J. D. Hooker (Nov. 3, 1880) shows the interest which he felt in the question:—

"I was delighted with Paget's Essay;* I hear that he has occasionally attended to this subject from his youth . . . . I am very glad he has called attention to galls: this has always seemed to me a profoundly interesting subject; and if I had been younger would take it up."

His interest in this subject was connected with his ever-present wish to learn something of the causes of variation. He imagined to himself wonderful galls caused to appear on the ovaries of plants, and by these means he thought it possible that the seed might be influenced, and thus new varieties arise. He made a considerable number of experiments by injecting various reagents into the tissues of leaves, and with some slight indications of success.]

AGGREGATION.

[The following letter gives an idea of the subject of the last of his published papers.† The appearances which he observed in leaves and roots attracted him, on account of their relation to the phenomena of aggregation which had so deeply interested him when he was at work on Drosera:]

C. Darwin to S. H. Vines.‡

Down, November 1, 1881.

MY DEAR MR. VINES,—As I know how busy you are, it is a great shame to trouble you. But you are so rich in

* 'Disease in Plants,' by Sir James Paget.—See *Gardeners' Chronicle*, 1880.

† 'Journal of the Linnean Society.' Vol. xix., 1882, pp. 239 and 262.

‡ Reader in Botany in the University of Cambridge.

chemical knowledge about plants, and I am so poor, that I appeal to your charity as a pauper. My question is—Do you know of any solid substance in the cells of plants which glycerine and water dissolves? But you will understand my perplexity better if I give you the facts: I mentioned to you that if a plant of *Euphorbia peplus* is gently dug up and the roots placed for a short time in a weak solution (1 to 10,000 of water suffices in 24 hours) of carbonate of ammonia the (generally) alternate longitudinal rows of cells in every rootlet, from the root-cap up to the very top of the root (but not as far as I have yet seen in the green stem) become filled with translucent, brownish grains of matter. These rounded grains often cohere and even become confluent. Pure phosphate and nitrate of ammonia produce (though more slowly) the same effect, as does pure carbonate of soda.

Now, if slices of root under a cover-glass are irrigated with glycerine and water, every one of the innumerable grains in the cells disappear after some hours. What am I to think of this? . . . .

Forgive me for bothering you to such an extent; but I must mention that if the roots are dipped in boiling water there is no deposition of matter, and carbonate of ammonia afterwards produces no effect. I should state that I now find that the granular matter is formed in the cells immediately beneath the thin epidermis, and a few other cells near the vascular tissue. If the granules consisted of living protoplasm (but I can see no traces of movement in them), then I should infer that the glycerine killed them and aggregation ceased with the diffusion of invisibly minute particles, for I have seen an analogous phenomenon in Drosera.

If you can aid me, pray do so, and anyhow forgive me.

Yours very sincerely,

CH. DARWIN.

Mr. Torbitt's Experiments on the Potato-Disease.

[Mr. James Torbitt, of Belfast, has been engaged for the last twelve years in the difficult undertaking, in which he has been to a large extent successful, of raising fungus-proof varieties of the potato. My father felt great interest in Mr. Torbitt's work, and corresponded with him from 1876 onwards. The following letter, giving a clear account of Mr. Torbitt's method and of my father's opinion of the probability of its success, was written with the idea that Government aid for the work might possibly be obtainable:]

C. Darwin to T. H. Farrer.

Down, March 2, 1878.

My dear Farrer,—Mr. Torbitt's plan of overcoming the potato-disease seems to me by far the best which has ever been suggested. It consists, as you know from his printed letter, of rearing a vast number of seedlings from cross-fertilised parents, exposing them to infection, ruthlessly destroying all that suffer, saving those which resist best, and repeating the process in successive seminal generations. My belief in the probability of good results from this process rests on the fact of all characters whatever occasionally varying. It is known, for instance, that certain species and varieties of the vine resist phylloxera better than others. Andrew Knight found one variety or species of the apple which was not in the least attacked by coccus, and another variety has been observed in South Australia. Certain varieties of the peach resist mildew, and several other such cases could be given. Therefore there is no great improbability in a new variety of potato arising which would resist the fungus completely, or at least much better than any existing variety. With respect to the cross-fertilisation of two distinct seedling plants, it has been ascertained that the offspring thus raised inherit much more vigorous constitutions and generally are more prolific than seedlings from self-fertilised parents. It is also probable that cross-fertilisation would be especially valuable in the case of the potato, as there is reason to believe that the flowers are seldom crossed by our native insects ; and some varieties are absolutely sterile unless fertilised with pollen from a distinct variety. There is some evidence that the good effects from a cross are transmitted for several generations ; it would not, therefore be necessary to cross-fertilise the seedlings in each generation, though this would be desirable, as it is almost certain that a greater number of seeds would thus be obtained. It should be remembered that a cross between plants raised from the tubers of the same plant, though growing on distinct roots, does no more good than a cross between flowers on the same individual. Considering the whole subject, it appears to me that it would be a national misfortune if the cross-fertilised seeds in Mr. Torbitt's possession produced by parents which have already shown some power of resisting the disease, are not utilised by the Government, or some public body, and the process of selection continued during several more generations.

Should the Agricultural Society undertake the work, Mr. Torbitt's knowledge gained by experience would be especially valuable ; and an outline of the plan is given in his printed letter. It would be necessary that all the tubers produced by each plant should be collected separately, and carefully examined in each succeeding generation.

It would be advisable that some kind of potato eminently liable to the disease should be planted in considerable numbers near the seedlings so as to infect them.

Altogether the trial would be one requiring much care and extreme patience, as I know from experience with analogous work, and it may be feared that it would be difficult to find any one who would pursue the experiment with sufficient energy. It seems, therefore, to me highly desirable that

Mr. Torbitt should be aided with some small grant so as to continue the work himself.

Judging from his reports, his efforts have already been crowned in so short a time with more success than could have been anticipated; and I think you will agree with me, that any one who raises a fungus-proof potato will be a public benefactor of no common kind.

My dear Farrer, yours sincerely,

CHARLES DARWIN.

[After further consultation with Sir Thomas Farrer and with Mr. Caird, my father became convinced that it was hopeless to attempt to obtain Government aid. He wrote to Mr. Torbitt to this effect, adding, "it would be less trouble to get up a subscription from a few rich leading agriculturists than from Government. This plan I think you cannot object to, as you have asked nothing, and will have nothing whatever to do with the subscription. In fact, the affair is, in my opinion, a compliment to you." The idea thus broached was carried out, and Mr. Torbitt was enabled to continue his work by the aid of a sum to which Sir T. Farrer, Mr. Caird, my father, and a few friends, subscribed.

My father's sympathy and encouragement were highly valued by Mr. Torbitt, who tells me that without them he should long ago have given up his attempt. A few extracts will illustrate his fellow-feeling with Mr. Torbitt's energy and perseverance:—

"I admire your indomitable spirit. If any one ever deserved success, you do so, and I keep to my original opinion that you have a very good chance of raising a fungus-proof variety of the potato.

"A pioneer in a new undertaking is sure to meet with many disappointments, so I hope that you will keep up your courage, though we have done so very little for you."

Mr. Torbitt tells me that he still (1887) succeeds in raising varieties possessing well-marked powers of resisting disease; but this immunity is not permanent, and, after some years, the varieties become liable to the attacks of the fungus.]

The Kew Index of Plant-Names, or 'Nomenclator Botanicus Darwinianus'.

[Some account of my father's connection with the Index of Plant-names now (1887) in course of preparation at Kew will be found in Mr. B. Daydon Jackson's paper in the 'Journal of Botany,' 1887, p. 151. Mr. Jackson quotes the following statement by Sir J. D. Hooker:—

"Shortly before his death, Mr. Charles Darwin informed Sir Joseph Hooker that it was his intention to devote a considerable sum of money annually for some years in aid or furtherance of some work or works of practical utility to biological science, and to make provisions in his will in the event of these not being completed during his lifetime.

"Amongst other objects connected with botanical science, Mr. Darwin regarded with especial interest the importance of a complete index to the names and authors of the genera and species of plants known to botanists, together with their native countries. Steudel's 'Nomenclator' is the only existing work of this nature, and although now nearly half a century old, Mr. Darwin had found it of great aid in his own researches. It has been indispensable to every botanical institution, whether as a list of all known flowering plants, as an indication of their authors, or as a digest of botanical geography."

Since 1840, when the 'Nomenclator' was published, the number of described plants may be said to have doubled, so that the 'Nomenclator' is now seriously below the requirements of botanical work. To remedy this want, the 'Nomenclator' has been from time to time posted up in an interleaved copy in the Herbarium at Kew, by the help of "funds supplied by private liberality." *

My father, like other botanists, had as Sir Joseph Hooker points out, experienced the value of Steudel's work. He obtained plants from all sorts of sources, which were often incorrectly named, and he felt the necessity of adhering to the accepted nomenclature, so that he might convey to other workers precise indications as to the plants which he had studied. It was also frequently a matter of importance to him to know the native country of his experimental plants. Thus it was natural that he should recognize the desirability of completing and publishing the interleaved volume at Kew. The wish to help in this object was heightened by the admiration he felt for the results for which the world has to thank the Royal Gardens at Kew, and by his gratitude for the invaluable aid which for so many years he received from its Director and his staff. He expressly stated that it was his wish "to aid in some way the scientific work carried on at the Royal Gardens" †—which induced him to offer to supply funds for the completion of the Kew 'Nomenclator.'

The following passage, for which I am indebted to Professor Judd, is of interest, as illustrating the motives that actuated my father in this matter. Professor Judd writes:—

"On the occasion of my last visit to him, he told me that his income having recently greatly increased, while his wants remained the same, he was most anxious to devote what he could spare to the advancement of Geology or Biology. He dwelt in the most touching manner on the fact that he owed so much happiness and fame to the natural-history sciences

* Kew Gardens Report, 1881, p. 62.

† See 'Nature,' January 5, 1882.

which had been the solace of what might have been a painful existence;—and he begged me, if I knew of any research which could be aided by a grant of a few hundreds of pounds, to let him know, as it would be a delight to him to feel that he was helping in promoting the progress of science. He informed me at the same time that he was making the same suggestion to Sir Joseph Hooker and Professor Huxley with respect to Botany and Zoology respectively. I was much impressed by the earnestness, and, indeed, deep emotion, with which he spoke of his indebtedness to Science, and his desire to promote its interests."

Sir Joseph Hooker was asked by my father "to take into consideration, with the aid of the botanical staff at Kew and the late Mr. Bentham, the extent and scope of the proposed work, and to suggest the best means of having it executed. In doing this, Sir Joseph had further the advantage of the great knowledge and experience of Professor Asa Gray, of Cambridge, U.S.A., and of Mr. John Ball, F.R.S." *

The plan of the proposed work having been carefully considered, Sir Joseph Hooker was able to confide its elaboration in detail to Mr. B. Daydon Jackson, Secretary of the Linnean Society, whose extensive knowledge of botanical literature qualifies him for the task. My father's original idea of producing a modern edition of Steudel's 'Nomenclator' has been practically abandoned, the aim now kept in view is rather to construct a list of genera and species (with references) founded on Bentham and Hooker's 'Genera Plantarum.' The colossal nature of the work in progress at Kew may be estimated by the fact that the manuscript of the 'Index' is at the present time (1887) believed to weigh more than a ton. Under Sir Joseph Hooker's supervision the work goes steadily forward, being carried out with admirable zeal by Mr. Jackson, who devotes himself unsparingly to the enterprise, in which,

* 'Journal of Botany,' *loc. cit.*

too, he has the advantage of the interest in the work felt by Professor Oliver and Mr. Thiselton Dyer.

The Kew 'Index,' which will, in all probability, be ready to go to press in four or five years, will be a fitting memorial of my father: and his share in its completion illustrates a part of his character—his ready sympathy with work outside his own lines of investigation—and his respect for minute and patient labour in all branches of science.]

CHAPTER XIII.

CONCLUSION.

SOME idea of the general course of my father's health may have been gathered from the letters given in the preceding pages. The subject of health appears more prominently than is often necessary in a Biography, because it was, unfortunately, so real an element in determining the outward form of his life.

During the last ten years of his life the state of his health was a cause of satisfaction and hope to his family. His condition showed signs of amendment in several particulars. He suffered less distress and discomfort, and was able to work more steadily. Something has been already said of Dr. Bence Jones's treatment, from which my father certainly derived benefit. In later years he became a patient of Sir Andrew Clark, under whose care he improved greatly in general health. It was not only for his generously rendered service that my father felt a debt of gratitude towards Sir Andrew Clark. He owed to his cheering personal influence an often-repeated encouragement, which latterly added something real to his happiness, and he found sincere pleasure in Sir Andrew's friendship and kindness towards himself and his children.

Scattered through the past pages are one or two references to pain or uneasiness felt in the region of the heart. How far these indicate that the heart was affected early in life, I cannot pretend to say; in any case it is certain that he had no serious or permanent trouble of this nature until shortly before his death. In spite of the general improvement in his health, which has been above alluded to, there was a certain loss of physical vigour occasionally apparent during the last few years of his life. This is illustrated by a sentence in a letter to his old friend Sir James Sulivan, written on January 10, 1879: "My scientific work tires me more than it used to do, but I have nothing else to do, and whether one is worn out a year or two sooner or later signifies but little."

A similar feeling is shown in a letter to Sir J. D. Hooker of June 15, 1881. My father was staying at Patterdale, and wrote: "I am rather despondent about myself . . . . I have not the heart or strength to begin any investigation lasting years, which is the only thing which I enjoy, and I have no little jobs which I can do."

In July, 1881, he wrote to Mr. Wallace, "We have just returned home after spending five weeks on Ullswater; the scenery is quite charming, but I cannot walk, and everything tires me, even seeing scenery . . . . What I shall do with my few remaining years of life I can hardly tell. I have everything to make me happy and contented, but life has become very wearisome to me." He was, however, able to do a good deal of work, and that of a trying sort,* during the autumn of 1881, but towards the end of the year he was clearly in need of rest; and during the winter was in a lower condition than was usual with him.

On December 13, he went for a week to his daughter's house in Bryanston Street. During his stay in London he went to call on Mr. Romanes, and was seized when on the door-step with an attack apparently of the same kind as those which afterwards became so frequent. The rest of the incident, which I give in Mr. Romanes' words, is interesting too from a different point of view, as giving one more illustration of my father's scrupulous consideration for others:—

* On the action of carbonate of ammonia on roots and leaves.

"I happened to be out, but my butler, observing that Mr. Darwin was ill, asked him to come in. He said he would prefer going home, and although the butler urged him to wait at least until a cab could be fetched, he said he would rather not give so much trouble. For the same reason he refused to allow the butler to accompany him. Accordingly he watched him walking with difficulty towards the direction in which cabs were to be met with, and saw that, when he had got about three hundred yards from the house, he staggered and caught hold of the park-railings as if to prevent himself from falling. The butler therefore hastened to his assistance, but after a few seconds saw him turn round with the evident purpose of retracing his steps to my house. However, after he had returned part of the way he seems to have felt better, for he again changed his mind, and proceeded to find a cab."

During the last week of February and in the beginning of March, attacks of pain in the region of the heart, with irregularity of the pulse, became frequent, coming on indeed nearly every afternoon. A seizure of this sort occurred about March 7, when he was walking alone at a short distance from the house; he got home with difficulty, and this was the last time that he was able to reach his favourite 'Sand-walk.' Shortly after this, his illness became obviously more serious and alarming, and he was seen by Sir Andrew Clark, whose treatment was continued by Dr. Norman Moore, of St. Bartholomew's Hospital, and Mr. Allfrey, of St. Mary Cray. He suffered from distressing sensations of exhaustion and faintness, and seemed to recognise with deep depression the fact that his working days were over. He gradually recovered from this condition, and became more cheerful and hopeful, as is shown in the following letter to Mr. Huxley, who was anxious that my father should have closer medical supervision than the existing arrangements allowed :—

Down, March 27, 1882.

"MY DEAR HUXLEY,—Your most kind letter has been a real cordial to me. I have felt better to-day than for three weeks, and have felt as yet no pain. Your plan seems an excellent one, and I will probably act upon it, unless I get very much better. Dr. Clark's kindness is unbounded to me, but he is too busy to come here. Once again, accept my cordial thanks, my dear old friend. I wish to God there were more automata * in the world like you.

Ever yours,
CH. DARWIN."

The allusion to Sir Andrew Clark requires a word of explanation. Sir Andrew Clark himself was ever ready to devote himself to my father, who, however, could not endure the thought of sending for him, knowing how severely his great practice taxed his strength.

No especial change occurred during the beginning of April, but on Saturday 15th he was seized with giddiness while sitting at dinner in the evening, and fainted in an attempt to reach his sofa. On the 17th he was again better, and in my temporary absence recorded for me the progress of an experiment in which I was engaged. During the night of April 18th, about a quarter to twelve, he had a severe attack and passed into a faint, from which he was brought back to consciousness with great difficulty. He seemed to recognise the approach of death, and said, "I am not the least afraid to die." All the next morning he suffered from terrible nausea and faintness, and hardly rallied before the end came.

He died at about four o'clock on Wednesday, April 19th, 1882.

* The allusion is to Mr. Huxley's address, "On the hypothesis that animals are automata, and its history," given at the Belfast Meeting of the British Association, 1874, and republished in 'Science and Culture.'

I close the record of my father's life with a few words of retrospect added to the manuscript of his 'Autobiography' in 1879:—

"As for myself, I believe that I have acted rightly in steadily following and devoting my life to Science. I feel no remorse from having committed any great sin, but have often and often regretted that I have not done more direct good to my fellow creatures."

THE END.

APPENDIX I.

THE FUNERAL IN WESTMINSTER ABBEY.

ON the Friday succeeding my father's death, the following letter, signed by twenty Members of Parliament, was addressed to Dr. Bradley, Dean of Westminster:—

HOUSE OF COMMONS, April 21, 1882.

VERY REV. SIR,—We hope you will not think we are taking a liberty if we venture to suggest that it would be acceptable to a very large number of our fellow-countrymen of all classes and opinions that our illustrious countryman, Mr. Darwin, should be buried in Westminster Abbey.

We remain your obedient servants,

JOHN LUBBOCK,
NEVIL STOREY MASKELYNE,
A. J. MUNDELLA,
G. O. TREVELYAN,
LYON PLAYFAIR,
CHARLES W. DILKE,
DAVID WEDDERBURN,
ARTHUR RUSSELL,
HORACE DAVEY,
BENJAMIN ARMITAGE,
RICHARD B. MARTIN,
FRANCIS W. BUXTON,
E. L. STANLEY,
HENRY BROADHURST,
JOHN BARRAN,
J. F. CHEETHAM,
H. S. HOLLAND,
H. CAMPBELL-BANNERMAN,
CHARLES BRUCE,
RICHARD FORT.

The Dean was abroad at the time, and telegraphed his cordial acquiescence.

The family had desired that my father should be buried at Down: with regard to their wishes, Sir John Lubbock wrote:—

House of Commons, April 25, 1882.

My dear Darwin,—I quite sympathise with your feeling, and personally I should greatly have preferred that your father should have rested in Down amongst us all. It is, I am sure, quite understood that the initiative was not taken by you. Still, from a national point of view, it is clearly right that he should be buried in the Abbey. I esteem it a great privilege to be allowed to accompany my dear master to the grave.

Believe me, yours most sincerely,

John Lubbock.

W. E. Darwin, Esq.

The family gave up their first-formed plans, and the funeral took place in Westminster Abbey on April 26th. The pall-bearers were :—

Sir John Lubbock,
Mr. Huxley,
Mr. James Russell Lowell (American Minister),
Mr. A. R. Wallace,
The Duke of Devonshire,
Canon Farrar,
Sir Joseph Hooker,
Mr. Wm. Spottiswoode (President of the Royal Society),
The Earl of Derby,
The Duke of Argyll.

The funeral was attended by the representatives of France, Germany, Italy, Spain, Russia, and by those of the Universities and learned Societies, as well as by large numbers of personal friends and distinguished men.

The grave is in the north aisle of the Nave, close to the angle of the choir-screen, and a few feet from the grave of Sir Isaac Newton. The stone bears the inscription—

CHARLES ROBERT DARWIN.

Born 12 February, 1809.

Died 19 April, 1882.

APPENDIX II.

I.—List of Works by C. Darwin.

Narrative of the Surveying Voyages of Her Majesty's Ships 'Adventure' and 'Beagle' between the years 1826 and 1836, describing their examination of the Southern shores of South America, and the 'Beagle's' circumnavigation of the globe. Vol. iii. Journal and Remarks, 1832–1836. By Charles Darwin. 8vo. London, 1839.

Journal of Researches into the Natural History and Geology of the countries visited during the Voyage of H.M.S. 'Beagle' round the world, under the command of Capt. Fitz-Roy, R.N. 2nd edition, corrected, with additions. 8vo. London, 1845. (Colonial and Home Library.)

A Naturalist's Voyage. Journal of Researches, &c. 8vo. London, 1860. [Contains a postscript dated Feb. 1, 1860.]

Zoology of the Voyage of H.M.S. 'Beagle.' Edited and superintended by Charles Darwin. Part I. Fossil Mammalia, by Richard Owen. With a Geological Introduction, by Charles Darwin. 4to. London, 1840.

—— Part II. Mammalia, by George R. Waterhouse. With a notice of their habits and ranges, by Charles Darwin. 4to. London, 1839.

—— Part III. Birds, by John Gould. An "Advertisement" (2 pp.) states that in consequence of Mr. Gould's having left England for Australia, many descriptions were supplied by Mr. G. R. Gray of the British Museum. 4to. London, 1841.

—— Part IV. Fish, by Rev. Leonard Jenyns. 4to. London, 1842.

—— Part V. Reptiles, by Thomas Bell. 4to. London, 1843.

The Structure and Distribution of Coral Reefs. Being the First

Part of the Geology of the Voyage of the 'Beagle.' 8vo. London, 1842.

The Structure and Distribution of Coral Reefs. 2nd edition. 8vo. London, 1874.

Geological Observations on the Volcanic Islands, visited during the Voyage of H.M.S. 'Beagle.' Being the Second Part of the Geology of the Voyage of the 'Beagle.' 8vo. London, 1844.

Geological Observations on South America. Being the Third Part of the Geology of the Voyage of the 'Beagle.' 8vo. London, 1846.

Geological Observations on the Volcanic Islands and parts of South America visited during the Voyage of H.M.S. 'Beagle.' 2nd edition. 8vo. London, 1876.

A Monograph of the Fossil Lepadidæ; or, Pedunculated Cirripedes of Great Britain. 4to. London, 1851. (Palæontographical Society.)

A Monograph of the Sub-class Cirripedia, with Figures of all the Species. The Lepadidæ; or, Pedunculated Cirripedes. 8vo. London, 1851. (Ray Society.)

—— The Balanidæ (or Sessile Cirripedes); the Verrucidæ, &c. 8vo. London, 1854. (Ray Society.)

A Monograph of the Fossil Balanidæ and Verrucidæ of Great Britain. 4to. London, 1854. (Palæontographical Society.)

On the Origin of Species by means of Natural Selection, or the Preservation of Favoured Races in the Struggle for Life. 8vo. London, 1859. (Dated Oct. 1st, 1859, published Nov. 24, 1859.)

—— Fifth thousand. 8vo. London, 1860.

—— Third edition, with additions and corrections. (Seventh thousand.) 8vo. London, 1861. (Dated March, 1861.)

—— Fourth edition, with additions and corrections. (Eighth thousand.) 8vo. London, 1866. (Dated June, 1866.)

—— Fifth edition, with additions and corrections. (Tenth thousand.) 8vo. London, 1869. (Dated May, 1869.)

—— Sixth edition, with additions and corrections to 1872. (Twenty-fourth thousand.) 8vo. London, 1882. (Dated Jan., 1872.)

On the various contrivances by which Orchids are fertilised by Insects. 8vo. London, 1862.

—— Second edition. 8vo. London, 1877. [In the second edition the word "On" is omitted from the title.]

The Movements and Habits of Climbing Plants. Second edition. 8vo. London, 1875. [First appeared in the ninth volume of the 'Journal of the Linnean Society.']

The Variation of Animals and Plants under Domestication. 2 vols. 8vo. London, 1868.

—— Second edition, revised. 2 vols. 8vo. London, 1875.

The Descent of Man, and Selection in Relation to Sex. 2 vols. 8vo. London, 1871.

—— Second edition. 8vo. London, 1874. (In 1 vol.)

The Expression of the Emotions in Man and Animals. 8vo. London, 1872.

The Effects of Cross and Self Fertilisation in the Vegetable Kingdom. 8vo. London, 1876.

—— Second edition. 8vo. London, 1878.

The different Forms of Flowers on Plants of the same Species. 8vo. London, 1877.

—— Second edition. 8vo. London, 1880.

The Power of Movement in Plants. By Charles Darwin, assisted by Francis Darwin. 8vo. London, 1880.

The Formation of Vegetable Mould, through the Action of Worms, with Observations on their Habits. 8vo. London, 1881.

II.—List of Books containing Contributions by C. Darwin.

A manual of scientific enquiry; prepared for the use of Her Majesty's Navy: and adapted for travellers in general. Ed. by Sir John F. W. Herschel, Bart. 8vo. London, 1849. (Section VI. Geology. By Charles Darwin.)

Memoir of the Rev. John Stevens Henslow. By the Rev. Leonard Jenyns. 8vo. London, 1862. [In Chapter III., Recollections by C. Darwin.]

A letter (1876) on the 'Drift' near Southampton, published in Prof. J. Geikie's 'Prehistoric Europe.'

Flowers and their unbidden guests. By A. Kerner. With a Prefatory Letter by Charles Darwin. The translation revised and edited by W. Ogle. 8vo. London, 1878.

Erasmus Darwin. By Ernst Krause. Translated from the German by W. S. Dallas. With a preliminary notice by Charles Darwin. 8vo. London, 1879.

Studies in the Theory of Descent. By Aug. Weismann. Translated and edited by Raphael Meldola. With a Prefatory Notice by Charles Darwin. 8vo. London, 1880—.

The Fertilisation of Flowers. By Hermann Müller. Translated and edited by D'Arcy W. Thompson. With a Preface by Charles Darwin. 8vo. London, 1883.

Mental Evolution in Animals. By G. J. Romanes. With a posthumous essay on instinct by Charles Darwin, 1883. [Also published in the Journal of the Linnean Society.]

Some Notes on a curious habit of male humble bees were sent to Prof. Hermann Müller, of Lippstadt, who had permission from Mr. Darwin to make what use he pleased of them. After Müller's death the Notes were given by his son to Dr. E. Krause, who published them under the title, "Ueber die Wege der Hummel-Männchen" in his book, 'Gesammelte kleinere Schriften von Charles Darwin' (1887).

III.—List of Scientific Papers, including a selection of Letters and Short Communications to Scientific Journals.

Letters to Professor Henslow, read by him at the meeting of the Cambridge Philosophical Society, held Nov. 16, 1835. 31 pp. 8vo. Privately printed for distribution among the members of the Society.

Geological Notes made during a survey of the East and West Coasts of South America in the years 1832, 1833, 1834, and 1835; with an account of a transverse section of the Cordilleras of the Andes between Valparaiso and Mendoza. [Read Nov. 18, 1835.] Geol. Soc. Proc. ii. 1838, pp. 210–212. [This Paper is incorrectly described in Geol. Soc. Proc. ii., p. 210 as follows:—"Geological notes, &c., by F. Darwin, Esq., of St. John's College, Cambridge: communicated by Prof. Sedgwick." It is Indexed under C. Darwin.]

Notes upon the Rhea Americana. Zool. Soc. Proc., Part v. 1837, pp. 35–36.

Observations of proofs of recent elevation on the coast of Chili, made during the survey of H.M.S. "Beagle," commanded by Capt. FitzRoy. [1837.] Geol. Soc. Proc. ii. 1838, pp. 446–449.

A sketch of the deposits containing extinct Mammalia in the neighbourhood of the Plata. [1837.] Geol. Soc. Proc. ii. 1838, pp. 542–544.

On certain areas of elevation and subsidence in the Pacific and

Indian oceans, as deduced from the study of coral formations. [1837.] Geol. Soc. Proc. ii. 1838, pp. 552–554.

On the Formation of Mould. [Read Nov. 1, 1837.] Geol. Soc. Proc. ii. 1838, pp. 574–576; Geol. Soc. Trans. v. 1840, pp. 505–510.

On the Connexion of certain Volcanic Phenomena and on the formation of mountain-chains and the effects of continental elevations. [Read March 7, 1838.] Geol. Soc. Proc. ii. 1838, pp. 654–660; Geol. Soc. Trans. v. 1840, pp. 601–632. [In the Society's Transactions the wording of the title is slightly different.]

Origin of saliferous deposits. Salt Lakes of Patagonia and La Plata. Geol. Soc. Journ. ii. (Part ii.), 1838, pp. 127–128.

Note on a Rock seen on an Iceberg in 16° South Latitude. Geogr. Soc. Journ. ix. 1839, pp. 528–529.

Observations on the Parallel Roads of Glen Roy, and of other parts of Lochaber in Scotland, with an attempt to prove that they are of marine origin. Phil. Trans. 1839, pp. 39–82.

On a remarkable Bar of Sandstone off Pernambuco, on the Coast of Brazil. Phil. Mag. xix. 1841, pp. 257–260.

On the Distribution of the Erratic Boulders and on the Contemporaneous Unstratified Deposits of South America. [1841.] Geol. Soc. Proc. iii. 1842, pp. 425–430; Geol. Soc. Trans. [1841.] vi. 1842, pp. 415–432.

Notes on the Effects produced by the Ancient Glaciers of Caernarvonshire, and on the Boulders transported by Floating Ice. London Philosoph. Mag. vol. xxi. p. 180. 1842.

Remarks on the preceding paper, in a Letter from Charles Darwin, Esq., to Mr. Maclaren. Edinb. New Phil. Journ. xxxiv. 1843, pp. 47–50. [The "preceding" paper is: "On Coral Islands and Reefs as described by Mr. Darwin. By Charles Maclaren, Esq., F.R.S.E."]

Observations on the Structure and Propagation of the genus *Sagitta*. Ann. and Mag. Nat. Hist. xiii. 1844, pp. 1–6.

Brief Descriptions of several Terrestrial *Planariæ*, and of some remarkable Marine Species, with an Account of their Habits. Ann. and Mag. Nat. Hist. xiv. 1844, pp. 241–251.

An account of the Fine Dust which often falls on Vessels in the Atlantic Ocean. Geol. Soc. Journ. ii. 1846, pp. 26–30.

On the Geology of the Falkland Islands. Geol. Soc. Journ. ii. 1846, pp. 267–274.

A review of Waterhouse's 'Natural History of the Mammalia.' [Not signed.] Ann. and Mag. of Nat. Hist. 1847. Vol. xix. p. 53.

On the Transportal of Erratic Boulders from a lower to a higher level. Geol. Soc. Journ. iv. 1848, pp. 315–323.

On British fossil Lepadidæ. Geol. Soc. Journ. vi. 1850, pp. 439–440. [The G. S. J. says, "This paper was withdrawn by the author with the permission of the Council."]

Analogy of the Structure of some Volcanic Rocks with that of Glaciers. Edinb. Roy. Soc. Proc. ii. 1851, pp. 17–18.

On the power of Icebergs to make rectilinear, uniformly-directed Grooves across a Submarine Undulatory Surface. Phil. Mag. x. 1855, pp. 96–98.

Vitality of Seeds. *Gardeners' Chronicle*, Nov. 17, 1855, p. 758.

On the action of Sea-water on the Germination of Seeds. [1856.] Linn. Soc. Journ. i. 1857 (*Botany*), pp. 130–140.

On the Agency of Bees in the Fertilisation of Papilionaceous Flowers. *Gardeners' Chronicle*, p. 725, 1857.

On the Tendency of Species to form Varieties; and on the Perpetuation of Varieties and Species by Natural Means of Selection. By Charles Darwin, Esq., F.R.S., F.L.S., and F.G.S., and Alfred Wallace, Esq. [Read July 1st, 1858.] Journ. Linn. Soc. 1859, vol. iii. (*Zoology*), p. 45.

Special titles of C. Darwin's contributions to the foregoing:—

(i) Extract from an unpublished work on Species by C. Darwin, Esq., consisting of a portion of a chapter entitled, "On the Variation of Organic Beings in a State of Nature; on the Natural Means of Selection; on the Comparison of Domestic Races and true Species." (ii) Abstract of a Letter from C. Darwin, Esq., to Professor Asa Gray, of Boston, U.S., dated Sept. 5, 1857.

On the Agency of Bees in the Fertilization of Papilionaceous Flowers, and on the Crossing of Kidney Beans. *Gardeners' Chronicle*, 1858, p. 828 and Ann. Nat. Hist. 3rd series ii. 1858, pp. 459–465.

Do the Tineina or other small Moths suck Flowers, and if so what Flowers? *Entom. Weekly Intell.* vol. viii. 1860, p. 103.

Note on the achenia of *Pumilio Argyrolepis*. *Gardeners' Chronicle*, Jan. 5, 1861, p. 4.

Fertilisation of Vincas. *Gardeners' Chronicle*, pp. 552, 831, 832. 1861.

On the Two Forms, or Dimorphic Condition, in the species of

Primula, and on their remarkable Sexual Relations. Linn. Soc. Journ. vi. 1862 (*Botany*), pp. 77–96.

On the Three remarkable Sexual Forms of *Catasetum tridentatum*, an Orchid in the possession of the Linnean Society. Linn. Soc. Journ. vi. 1862 (*Botany*), pp. 151–157.

Yellow Rain. *Gardeners' Chronicle*, July 18, 1863, p. 675.

On the thickness of the Pampean formation near Buenos Ayres. Geol. Soc. Journ. xix. 1863, pp. 68–71.

On the so-called "Auditory-sac" of Cirripedes. Nat. Hist. Review, 1863, pp. 115–116.

A review of Mr. Bates' paper on 'Mimetic Butterflies.' Nat. Hist. Review, 1863, p. 221—. [Not signed.]

On the existence of two forms, and on their reciprocal sexual relation, in several species of the genus *Linum*. Linn. Soc. Journ. vii. 1864 (*Botany*), pp. 69–83.

On the Sexual Relations of the Three Forms of *Lythrum salicaria*. [1864.] Linn. Soc. Journ. viii. 1865 (*Botany*), pp. 169–196.

On the Movement and Habits of Climbing Plants. [1865.] Linn. Soc. Journ. ix. 1867 (*Botany*), pp. 1–118.

Note on the Common Broom (*Cytisus scoparius*). [1866.] Linn. Soc. Journ. ix. 1867 (*Botany*), p. 358.

Notes on the Fertilization of Orchids. Ann. and Mag. Nat. Hist. 4th series, iv. 1869, pp. 141–159.

On the Character and Hybrid-like Nature of the Offspring from the Illegitimate Unions of Dimorphic and Trimorphic Plants. [1868.] Linn. Soc. Jour. x. 1869 (*Botany*), pp. 393–437.

On the Specific Difference between *Primula veris*, Brit. Fl. (var. *officinalis*, of Linn.), *P. vulgaris*, Brit. Fl. (var. *acaulis*, Linn.), and *P. elatior*, Jacq.; and on the Hybrid Nature of the common Oxlip. With Supplementary Remarks on naturally-produced Hybrids in the genus *Verbascum*. [1868.] Linn. Soc. Journ. x. 1869 (*Botany*), pp. 437–454.

Note on the Habits of the Pampas Woodpecker (*Colaptes campestris*). Zool. Soc. Proc. Nov. 1, 1870, pp. 705–706.

Fertilisation of *Leschenaultia*. *Gardeners' Chronicle*, p. 1166, 1871.

The Fertilisation of Winter-flowering Plants. 'Nature,' Nov. 18, 1869, vol. i. p. 85.

Pangenesis. 'Nature,' April 27, 1871, vol. iii. p. 502.

A new view of Darwinism. 'Nature,' July 6, 1871, vol. iv. p. 180.

Bree on Darwinism. 'Nature,' Aug. 8, 1872, vol. vi. p. 279.

Inherited Instinct. 'Nature,' Feb. 13, 1873, vol. vii. p. 281.

Perception in the Lower Animals. 'Nature,' March 13, 1873, vol. vii. p. 360.

Origin of certain instincts. 'Nature,' April 3, 1873, vol. vii. p. 417.

Habits of Ants. 'Nature,' July 24, 1873, vol. viii. p. 244.

On the Males and Complemental Males of Certain Cirripedes, and on Rudimentary Structures. 'Nature,' Sept. 25, 1873, vol. viii. p. 431.

Recent researches on Termites and Honey-bees. 'Nature,' Feb. 19, 1874, vol. ix. p. 308.

Fertilisation of the Fumariaceæ. 'Nature,' April 16, 1874, vol. ix. p. 460.

Flowers of the Primrose destroyed by Birds. 'Nature,' April 23, 1874, vol. ix. p. 482; May 14, 1874, vol. x. p. 24.

Cherry Blossoms. 'Nature,' May 11, 1876, vol. xiv. p. 28.

Sexual Selection in relation to Monkeys. 'Nature,' Nov. 2, 1876, vol. xv. p. 18.

Fritz Müller on Flowers and Insects. 'Nature,' Nov. 29, 1877, vol. xvii. p. 78.

The Scarcity of Holly Berries and Bees. *Gardeners' Chronicle*, Jan. 20, 1877, p. 83.

Note on Fertilisation of Plants. *Gardeners' Chronicle*, vol. vii. p. 246, 1877.

A biographical sketch of an infant. 'Mind,' No. 7, July, 1877.

Transplantation of Shells. 'Nature,' May 30, 1878, vol. xviii. p. 120.

Fritz Müller on a Frog having Eggs on its back—on the abortion of the hairs on the legs of certain Caddis-Flies, &c. 'Nature,' March 20, 1879, vol. xix. p. 462.

Rats and Water-Casks. 'Nature,' March 27, 1879, vol. xix. p. 481.

Fertility of Hybrids from the common and Chinese Goose. 'Nature,' Jan. 1, 1880, vol. xxi. p. 207.

The Sexual Colours of certain Butterflies. 'Nature,' Jan. 8, 1880, vol. xxi. p. 237.

The Omori Shell Mounds. 'Nature,' April 15, 1880, vol. xxi. p. 561.

Sir Wyville Thomson and Natural Selection. 'Nature,' Nov. 11, 1880, vol. xxiii. p. 32.

Black Sheep. 'Nature,' Dec. 30, 1880, vol. xxiii. p. 193.

Movements of Plants. 'Nature,' March 3, 1881, vol. xxiii. p. 409.

The Movements of Leaves. 'Nature,' April 28, 1881, vol. xxiii. p. 603.

Inheritance. 'Nature,' July 21, 1881, vol. xxiv. p. 257.

Leaves injured at Night by Free Radiation. 'Nature,' Sept. 15, 1881, vol. xxiv. p. 459.

The Parasitic Habits of Molothrus. 'Nature,' Nov. 17, 1881, vol. xxv. p. 51.

On the Dispersal of Freshwater Bivalves. 'Nature,' April 6, 1882, vol. xxv. p. 529.

The Action of Carbonate of Ammonia on the Roots of certain Plants. [Read March 16, 1882.] Linn. Soc. Journ. (*Botany*), vol. xix. 1882, pp. 239–261.

The Action of Carbonate of Ammonia on Chlorophyll-bodies. [Read March 6, 1882.] Linn. Soc. Journ. (*Botany*), vol. xix. 1882, pp. 262–284.

On the Modification of a Race of Syrian Street-Dogs by means of Sexual Selection. By W. Van Dyck. With a preliminary notice by Charles Darwin. [Read April 18, 1882.] Proc. Zoolog. Soc. 1882, pp. 367–370.

APPENDIX III.

PORTRAITS.

Date.	Description.	Artist.	In the Possession of
1838	Water-colour .	G. Richmond .	The Family.
1851	Lithograph . .	Ipswich British Assn. Series.	
1853	Chalk Drawing .	Samuel Lawrence	The Family.
1853?	Chalk Drawing *	Samuel Lawrence	Prof. Hughes, Cambridge.
1869	Bust, marble .	T. Woolner, R.A.	The Family.
1875	Oil Painting † .	W. Ouless, R.A.	The Family.
	Etched by	P. Rajon.	
1879	Oil Painting .	W. B. Richmond	The University of Cambridge.
1881	Oil Painting ‡ .	Hon. John Collier	The Linnean Society.
	Etched by	Leopold Flameng	

CHIEF PORTRAITS AND MEMORIALS NOT TAKEN FROM LIFE

Date.	Description.	Artist.	In the Possession of
	Statue . . .	Joseph Boehm, R.A.	Museum, South Kensington.
	Bust . . .	Chr. Lehr, Junr.	
	Plaque . . .	T. Woolner, R.A., and Josiah Wedgwood and Sons.	Christ's College, in Charles Darwin's Room.
	Deep Medallion	J. Boehm, R.A.	To be placed in Westminster Abbey.

* Probably a sketch made at one of the sittings for the last-mentioned.

† A *replica* by the artist is in the possession of Christ's College, Cambridge.

‡ A *replica* by the artist is in the possession of W. E. Darwin, Esq., Southampton.

Chief Engravings from Photographs.

*1854? By Messrs. Maull and Fox, engraved on wood for 'Harper's Magazine' (Oct. 1884). Frontispiece, vol. i.

*1870? By O. J. Rejlander, engraved on steel by C. H. Jeens for 'Nature' (June 4, 1874).

*1874? By Capt. Darwin, R.E., engraved on wood for the 'Century Magazine' (Jan. 1883). Frontispiece, vol. ii.

1881 By Messrs. Elliott and Fry, engraved on wood by G. Kruells, for vol. iii. of the present work.

* The dates of these photographs must, from various causes, remain uncertain. Owing to a loss of books by fire, Messrs. Maull and Fox can give only an approximate date. Mr. Rejlander died some years ago, and his business was broken up. My brother, Captain Darwin, has no record of the date at which his photograph was taken.

APPENDIX IV.*

HONOURS, DEGREES, SOCIETIES, &c.

Order.—Prussian Order, 'Pour le Mérite.' 1867.

Office.—County Magistrate. 1857.

Degrees.—Cambridge { B.A. 1831 [1832].†
M.A. 1837.
Hon. LL.D. 1877.

Bonn . . Hon. Doctor in Medicine and Surgery. 1868.

Breslau . Hon. Doctor in Medicine and Surgery. 1862.

Leyden . Hon. M.D. 1875.

Societies.—London . Zoological. Corresp. Member. 1831.‡

Entomological. 1833, Orig. Member.

Geological. 1836. Wollaston Medal, 1859.

Royal Geographical. 1838.

Royal. 1839. Royal Society's Medal, 1853 Copley Medal, 1864.

Linnean. 1854.

Ethnological. 1861.

Medico-Chirurgical. Hon. Member. 1868.

Baly Medal of the Royal College of Physicians, 1879.

Societies.—Provincial, Colonial and Indian.

Royal Society of Edinburgh, 1865.

Royal Medical Society of Edinburgh, 1826. Hon. Member, 1861.

Royal Irish Academy. Hon. Member, 1866.

* The list has been compiled from the diplomas and letters in my father's possession, and is no doubt incomplete, as he seems to have lost or mislaid some of the papers received from foreign Societies. Where the name of a foreign Society (excluding those in the United States) is given in English, it is a translation of the Latin (or in one case Russian) of the original Diploma.

† See vol. i. p. 163.

‡ He afterwards became a Fellow of the Society.

Literary and Philosophical Society of Manchester. Hon. Member, 1868.

Watford Nat. Hist. Society. Hon. Member, 1877.

Asiatic Society of Bengal. Hon. Member, 1871.

Royal Society of New South Wales. Hon. Member, 1879.

Philosophical Institute of Canterbury, New Zealand. Hon. Member, 1863.

New Zealand Institute. Hon. Member, 1872.

Foreign Societies.

AMERICA.

Sociedad Científica Argentina. Hon. Member, 1877.

Academia Nacional de Ciencias, Argentine Republic. Hon. Member, 1878.

Sociedad Zoolójica Arjentina. Hon. Member, 1874.

Boston Society of Natural History. Hon. Member, 1873.

American Academy of Arts and Sciences (Boston). Foreign Hon. Member, 1874.

California Academy of Sciences. Hon. Member, 1872.

California State Geological Society. Corresp. Member, 1877.

Franklin Literary Society, Indiana. Hon. Member, 1878.

Sociedad de Naturalistas Neo-Granadinos. Hon. Member, 1860.

New York Academy of Sciences. Hon. Member, 1879.

Gabinete Portuguez de Leitura em Pernambuco. Corresp. Member, 1879.

Academy of Natural Sciences of Philadelphia. Correspondent, 1860.

American Philosophical Society, Philadelphia. Member, 1869.

AUSTRIA-HUNGARY.

Imperial Academy of Sciences of Vienna. Foreign Corresponding Member, 1871; Hon. Foreign Member, 1875.

Anthropologische Gesellschaft in Wien. Hon. Member, 1872.

K. k. Zoologische botanische Gesellschaft in Wien. Member, 1867.

Magyar Tudományos Akadémia, Pest, 1872.

BELGIUM.

Société Royale des Sciences Médicales et Naturelles de Bruxelles. Hon. Member, 1878.

Société Royale de Botanique de Belgique. 'Membre Associé,' 1881

Académie Royale des Sciences, &c., de Belgique. 'Associé de la Classe des Sciences.' 1870.

Denmark.

Royal Society of Copenhagen. Fellow, 1879.

France.

Société d'Anthropologie de Paris. Foreign Member, 1871.
Société Entomologique de France. Hon. Member, 1874.
Société Géologique de France. Life Member, 1837.
Institut de France. 'Correspondant' Section of Botany, 1878.

Germany.

Royal Prussian Academy of Sciences (Berlin). Corresponding Member, 1863; Fellow, 1878.
Berliner Gesellschaft für Anthropologie, &c. Corresponding Member, 1877.
Schlesische Gesellschaft für Vaterländische Cultur (Breslau). Hon. Member, 1878.
Cæsarea Leopoldino-Carolina Academia Naturæ Curiosorum (Dresden).* 1857.
Senkenbergische Naturforschende Gesellschaft zu Frankfurt am Main. Corresponding Member, 1873.
Naturforschende Gesellschaft zu Halle. Member, 1879.
Siebenbürgische Verein für Naturwissenschaften (Hermannstadt). Hon. Member, 1877.
Medicinisch-naturwissenschaftliche Gesellschaft zu Jena. Hon. Member, 1878.
Royal Bavarian Academy of Literature and Science (Munich). Foreign Member, 1878.

Holland.

Koninklijke Natuurkundige Vereeniging in Nederlandsch-Indie (Batavia). Corresponding Member, 1880.

* The diploma contains the words "accipe . . . ex antiqua nostra consuetudine cognomen Forster." It was formerly the custom in the *Cæsarea Leopoldino-Carolina Academia*, that each new member should receive as a 'cognomen,' a name celebrated in that branch of science to which he belonged. Thus a physician might be christened Boerhaave, or an astronomer, Kepler. My father seems to have been named after the traveller John Reinhold Forster.

Société Hollandaise des Sciences à Harlem. Foreign Member, 1877.

Zeeuwsch Genootschap der Wetenschappen te Middelburg. Foreign Member, 1877.

ITALY.

Società Geografica Italiana (Florence). 1870.

Società Italiana di Antropologia e di Etnologia (Florence). Hon. Member, 1872.

Società dei Naturalisti in Modena. Hon. Member, 1875.

Academia de' Lincei di Roma. Foreign Member, 1875.

La Scuola Italica, Academia Pitagorica, Reale ed Imp. Società (Rome). 'Presidente Onorario degli Anziani Pitagorici,' 1880.

Royal Academy of Turin. 1873. *Bressa* Prize, 1879.

PORTUGAL.

Sociedade de Geographia de Lisboa (Lisbon). Corresponding Member, 1877.

RUSSIA.

Society of Naturalists of the Imperial Kazan University. Hon. Member, 1875.

Societas Cæsarea Naturæ Curiosorum (Moscow). Hon. Member, 1870.

Imperial Academy of Sciences (St. Petersburg). Corresponding Member, 1867.

SPAIN.

Institucion Libre de Enseñanza (Madrid). Hon. Professor, 1877.

SWEDEN.

Royal Swedish Acad. of Sciences (Stockholm). Foreign Member, 1865.

Royal Society of Sciences (Upsala). Fellow, 1860.

SWITZERLAND.

Société des Sciences Naturelles du Neufchâtel. Corresponding Member, 1863.

INDEX.

Printed in the United States
By Bookmasters